The Finite Element Method
in Heat Transfer Analysis

The Finite Element Method in Heat Transfer Analysis

R. W. Lewis

K. Morgan

University of Wales, Swansea
Wales, UK

H. R. Thomas

University of Wales, Cardiff
Wales, UK

K. N. Seetharamu

Indian Institute of Technology, Madras, India

JOHN WILEY & SONS

Chichester · New York · Brisbane : Toronto · Singapore

Other Wiley Editorial Offices

John Wiley & Sons, Inc., 605 Third Avenue,
New York, NY 10158-0012, USA

Jacaranda Wiley Ltd, G.P.O. Box 859, Brisbane,
Queensland 4001, Australia

John Wiley & Sons (Canada) Ltd, 22 Worcester Road,
Rexdale, Ontario M9W 1L1, Canada

John Wiley & Sons (SEA) Pte Ltd, 37 Jalan Pemimpin #05-04,
Block B, Union Industrial Building, Singapore 2057

British Library Cataloguing in Publication Data

A catalogue record for this book is available from the British Library

ISBN 0 471 93424 0 ; 0 471 94362 2

Typesetting by Thomson Press (India) Ltd., New Delhi

Contents

3   Time Stepping Methods for Heat Transfer   81

4   Non-linear Heat Conduction Analysis   99

5   Phase Change Problems—Solidification and Melting   123

6 Convective Heat Transfer   **162**

7 Further Developments   **191**

Preface

Over the past three decades, the use of numerical simulation with high-speed electronic computers has gained wide acceptance throughout most of the major branches of engineering. Our joint research work on the subject of the use of the finite element method in heat transfer analysis began in the mid 1970s and flourished in such a way that we decided to write this text.

The text has evolved in much the same way that our joint research did. First, we went through steady-state heat conduction problems, eventually increasing in complexity to include such difficulties as the non-linear analysis of coupled heat and mass transfer in capillary porous bodies and the calculation of shrinkage stresses using an elasto-plastic constitutive relationship. Our lectures at postgraduate level at our institutions and at finite element courses in different parts of the world have also been a stimulus to complete the text.

The need for the book stems from the fact that few texts of this kind exist despite a plethora of classical work on heat transfer problems. The topic is itself briefly covered in many texts on the finite element method, but the detail covered in this book has been available only in research papers up until now. There has been literally an explosion of research interest in the area and it has been impossible to reference every work since the early 1970s, as a spectrum of disciplines is involved.

The arrangement of the text proceeds in a logical order of complexity. The first chapter deals with the importance of heat transfer in engineering problems and derives the general heat conduction equation. Then the weak variational formulation and appropriate initial, boundary and interfacial conditions are discussed. In the second chapter, linear steady-state heat conduction problems are solved using the Galerkin form of the weak formulation. Initially the focus is on problems involving one space dimension in an attempt to develop concepts and to aid understanding of the process of matrix assembly. Solutions to two-dimensional problems are then given using rectangular elements. The solutions are worked for single and multiple elements to bring out the importance of the number of elements in the solution. Chapter 3 deals with the various time stepping schemes used in unsteady state problems, including topics such as stability and convergence of solutions. Chapter 4 deals with the solution of transient non-linear heat conduction problems and provides a number of examples. The problems of melting and solidification are dealt with at length in Chapter 5 in view of their practical importance in manufacturing processes like casting and welding, for example. The difficulties associated with the application of the finite element method to convection problems and the methods used to overcome them are given in Chapter 6, along with simple illustrative examples of convective heat transfer between two parallel heated plates.

In order to bring out the practical importance of heat transfer analysis in the engineering industry, Chapter 7 deals with the application of heat and mass transfer to drying problems and the calculation of both thermal and shrinkage stresses.

The preface would not be complete without acknowledging the continued support that we have received over the years from our respective research councils, industrial partners, The British Council, and various other funding sources, together with the invaluable assistance given by a large number of talented research associates, assistants and students with whom we have had the pleasure of working.

February 1995

1 Conduction Heat Transfer and Formulation

1.1 INTRODUCTION

There are many practical engineering problems that require the analysis of problems involving the transfer of heat. The solution of the equation of heat conduction is sufficient in many cases, while the area of application expands considerably if the equations governing coupled heat and mass transfer and/or thermal convection are considered. Combining a thermal analysis procedure with a thermal stress predictive capability can provide answers to questions of immediate concern in many industrial processes.

In the field of civil engineering, there are numerous examples of practical problems where the behaviour of the system under consideration may be predicted via a heat transfer analysis. The thermal properties of materials used for construction purposes often necessitate the use of some form of insulation to counter the effects of the wide range of temperatures produced by seasonal climatic changes. This is particularly true in highway construction, where it is important to predict the placement and quantity of insulating material in order to prevent possible damage to the road surface due to frost heave. At excavation sites, where the soil may be unstable because of friability or water saturation, artificial freezing of the soil is often employed to render it structurally stable. This typically involves the circulation of a refrigerated liquid through pipes sunk in concentric circular patterns around the excavation site. The high cost of the process, in terms of the refrigeration plant and energy consumption, means that a numerical simulation of the process can prove to be extremely useful. If the numerical method makes accurate predictions of the advance of the frozen region, the numerical results can be used to aid the design of the optimal arrangement of the cooling pipes and to predict the required running time necessary to produce an ice wall of the required thickness. In the foundry, it is imperative to understand the process of solidification in metal casting and the effect this produces on porosity formation. In high pressure die casting, it is often required to design a cooling system that will ensure the desired surface finish. Problems of this type can be investigated by a heat conduction analysis, taking account of latent heat effects and allowing for the variation in the thermal properties of the materials with temperature.

For the simulation of problems involving porous capillary materials, such as timber or ceramics, a coupled heat and mass transfer analysis is required. The optimal drying rate to ensure a reasonably stress-free end product will be tied in with the energy cost of the kiln schedule. Mathematical models, which can accurately predict the movement of moisture and heat in porous materials, are often the only means of gaining a better insight into the physical process. Similar models prove useful in designing geothermal energy extraction systems or in the analysis of advanced oil recovery by thermal methods. By increasing the complexity of the governing equations, it is possible to predict the aerodynamic heating of structures, such as re-entry vehicles, or of turbine blades in a jet engine. A related problem here is the computation of the thermally induced stresses and the prediction of the working life of the component. Such information is essential in the safety analysis of nuclear reactors and also in the modern steel industry, where the production rate of continuously cast steel is limited by the cracking induced by thermal stress development in the strand.

Exact analytical solutions of the governing equations of heat transfer can only be obtained for problems in which restrictive simplifying assumptions have been made with respect to geometry, material properties and boundary conditions. There is therefore no option but to turn to numerical solution methods for the analysis of practical problems, where such simplifications are not generally possible. The finite element method, with its flexibility in dealing with complex geometries, is an ideal approach to employ in the solution of such problems. However, the newcomer frequently finds the gap between finite element theory and practice quite daunting and it is the objective of this book to attempt to bridge this gap. The material included proceeds in an orderly fashion from the governing differential equations to the finite element formulation.

Initially, the basic differential equation governing heat conduction will be derived and the concept of a variational formulation of the problem will be introduced. The finite element method will then be used to generate a system of simultaneous equations which have to be solved to obtain the approximation to the temperature field in the body of interest. The process will be demonstrated in detail for the simple case of linear, steady-state problems in both one and two dimensions. For the solution of time-dependent (transient) problems, finite difference methods will be employed to determine the variation of the solution with time. Different time marching algorithms are presented and their relative merits discussed.

When the variation of the thermal properties of the materials are included in the mathematical model, the resulting equations are non-linear and techniques for dealing with this added complication are described. Some examples for which exact solutions are available are also given for one-, two- and three-dimensional problems. Solution of the non-linear problems can be extended to include the effects of latent heat addition or removal as in melting and solidification problems. These types of problems, along with some practical examples, are dealt with in the chapter on phase change. Convection heat transfer is important when the heat exchange is between a solid and a fluid. The solution methods for such problems and the difficulties encountered in the application of the standard Galerkin method are also dealt with. As has been noted above, there are many practical problems where the determination of the temperature (or moisture) distribution is only required as a prerequisite for the calculation of the resulting stress field. Methods for the determination of thermal or drying stresses in such problems will be detailed.

1.2 MODELLING OF HEAT CONDUCTION

1.2.1 Derivation of the governing equation

The equation governing the conduction of heat in a continuous medium can be derived by imposing the principle of conservation of heat energy over an arbitrary fixed volume, V, of the medium which is bounded by a closed surface S. For convenience the conservation statement is expressed in rate form and is written as:

rate of increase of heat in V = rate of heat conduction into V across S

$$+ \text{ rate of heat generation within } V \qquad (1.2.1)$$

If u denotes the specific internal energy of the medium, then

$$\text{rate of increase of heat in } V = \int_V \rho \frac{\partial u}{\partial t} \, dV \qquad (1.2.2)$$

where ρ is the density of the medium. Introducing the specific heat, c, of the medium defined by

$$c = \frac{du}{dT} \qquad (1.2.3)$$

where T is the temperature, means that we can write equation (1.2.2) as

$$\text{rate of increase of heat in } V = \int_V \rho c \frac{\partial T}{\partial t} \, dV \qquad (1.2.4)$$

To obtain an expression for the rate at which heat is conducted into V across S, we make use of Fourier's Law of Conduction. This is an empirical relationship which states that, for a surface with unit normal vector **n**, the rate at which heat is conducted across the surface, per unit area, in the direction of **n** is given by

$$q = - k(\text{grad } T) \cdot \mathbf{n} = - k \frac{\partial T}{\partial n} \qquad (1.2.5)$$

where k is a property of the medium termed the thermal conductivity. In this equation $\partial/\partial n$ denotes differentiation in the direction of $\mathbf{n}$ and q is termed the flux of heat in this direction. Thus, if $\mathbf{n}$ denotes the outward unit normal to S, it follows that

rate of heat conduction into V across S

$$= \int_S -q\,dS = \int_S k(\text{grad }T)\cdot\mathbf{n}\,dS = \int_V \text{div}\,(k\,\text{grad }T)\,dV \qquad (1.2.6)$$

where the Divergence Theorem has been applied. If it is assumed that heat generation in the medium is occurring at a rate Q per unit volume, then

rate of heat generation within $V = \int_V Q\,dV \qquad (1.2.7)$

Using equations (1.2.4), (1.2.6) and (1.2.7) in (1.2.1) produces the conservation statement:

$$\int_V \left(\rho c\,\frac{\partial T}{\partial t} - \text{div}\,(k\,\text{grad }T) - Q \right) dV = 0 \qquad (1.2.8)$$

and, since the volume V was arbitrarily chosen initially, it follows that

$$\rho c\,\frac{\partial T}{\partial t} = \text{div}\,(k\,\text{grad }T) + Q \qquad (1.2.9)$$

everywhere in the medium. This is the familiar form of the heat conduction equation for a non-stationary system.

If the medium is anisotropic, i.e. the conductivity depends upon the direction, the form of the heat conduction equation is modified to

$$\rho c\,\frac{\partial T}{\partial t} = \text{div}\,(\mathbf{k}\,\text{grad }T) + Q \qquad (1.2.10)$$

where

$$\mathbf{k} = \begin{bmatrix} k_{xx} & k_{xy} & k_{xz} \\ k_{yx} & k_{yy} & k_{yz} \\ k_{zx} & k_{zy} & k_{zz} \end{bmatrix} \qquad (1.2.11)$$

is a conductivity tensor and, for example, k_{xy} denotes the thermal conductivity in the x direction across a surface with normal in the y direction. If the conductivity k and the specific heat capacity ρc are assumed to be constant, and if the heat generation rate Q is independent of T, then equation (1.2.9) is linear and can be written as

$$\frac{1}{\alpha_1}\,\frac{\partial T}{\partial t} = \nabla^2 T + \frac{Q}{k} \qquad (1.2.12)$$

where $\alpha_1 = k/\rho c$ is termed the thermal diffusivity of the medium and ∇^2 denotes the Laplacian operator defined, in Cartesian coordinates, by

$$\nabla^2 = \frac{\partial^2}{\partial x^2} + \frac{\partial^2}{\partial y^2} + \frac{\partial^2}{\partial z^2} \qquad (1.2.13)$$

In the absence of heat generation within the medium, equation (1.2.12) reduces to the standard diffusion equation:

$$\frac{1}{\alpha_1} \frac{\partial T}{\partial t} = \nabla^2 T \qquad (1.2.14)$$

If, in addition, the temperature does not vary with time, steady-state conditions are said to exist and, in this case, the governing equation simplifies further to

$$\nabla^2 T = 0 \qquad (1.2.15)$$

which is just the Laplace equation.

1.2.2 Initial and boundary conditions

Suppose that the solution of the heat conduction equation (1.2.9) is required over an arbitrary domain Ω bounded by a closed surface, Γ, as illustrated in Figure 1.2.1. If the problem being modelled is independent of time (i.e. steady), the solution will be uniquely defined provided that we are able to supply appropriate boundary conditions. For the steady heat conduction equation, one condition has to be specified at each point of the boundary curve Γ and typical conditions of practical interest would be:

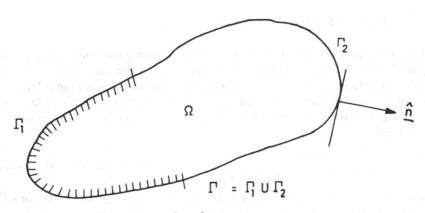

Figure 1.2.1 General domain and boundary.

(a) the value of the temperature is prescribed, e.g.

$$T = f(\mathbf{x}) \quad \text{for all } \mathbf{x} \text{ on } \Gamma_1 \tag{1.2.16}$$

or

(b) the value of the outward normal heat flux is prescribed, e.g.

$$q = -k(\text{grad } T) \cdot \mathbf{n} = -k\frac{\partial T}{\partial n} = \aleph(\mathbf{x}, T) + \aleph_c(\mathbf{x}, T) + \aleph_r(\mathbf{x}, T) \quad \text{for all } \mathbf{x} \text{ on } \Gamma_2$$

$$\tag{1.2.17}$$

Here $f, \aleph, \aleph_c$ and $\aleph_r$ are prescribed functions of $\mathbf{x}$ and T and $\Gamma = \Gamma_1 \cup \Gamma_2$, $\Gamma_1 \cap \Gamma_2 = 0$. In equation (1.2.17) $\aleph$ denotes a specified heat flux; $\aleph_c$ denotes a convective heat flux, defined as

$$\aleph_c = \alpha(T - T_\infty) \tag{1.2.18}$$

where α is a coefficient of surface heat transfer and T_∞ is the specified ambient temperature of the surrounding medium; $\aleph_r$ is a radiative heat flux which is defined as

$$\aleph_r = \varepsilon\sigma(T^4 - T_\infty^4) \tag{1.2.19}$$

where σ is the Stefan–Boltzman constant and ε is the emissivity of the surface, defined as the ratio of the heat emitted by the surface to the heat emitted by a black body at the same temperature.

When the problem being modelled is time dependent (transient), the solution is uniquely determined provided that an initial condition is given together with a boundary condition at each point of the boundary Γ of the domain. The initial condition should give the distribution of the temperature over the entire region Ω at an initial time, usually taken to be the time $t = 0$. In addition, in a transient problem, the functions $f, \aleph, \aleph_c$ and $\aleph_r$ of equations (1.2.16) and (1.2.17) may vary with time.

1.2.3 Interface conditions

When the region Ω, consists of two or more different materials, as shown in Figure 1.2.2, boundary conditions of continuity of temperature and flux across material interfaces have to be applied in order to uniquely define the solution. If we consider an interface $\tilde{\Gamma}$ between two materials, designated Ω_1 and Ω_2, with unit normal vector $\tilde{\mathbf{n}}$ in the direction into Ω_1, these conditions can be written in the form

$$\left. \begin{array}{l} T_1(\mathbf{x}, t) = T_2(\mathbf{x}, t) \\ \tilde{\mathbf{n}} \cdot (k \, \text{grad } T)_1 = \tilde{\mathbf{n}} \cdot (k \, \text{grad } T)_2 \end{array} \right\} \quad \text{for all } \mathbf{x} \text{ on } \tilde{\Gamma} \text{ and all } t > 0 \tag{1.2.20}$$

where the subscripts 1 and 2 denote the conditions appropriate to regions Ω_1 and Ω_2, respectively.

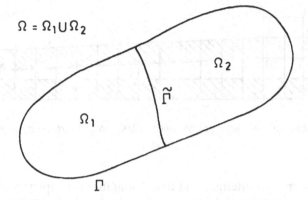

$\Omega = \Omega_1 \cup \Omega_2$

Ω_2

$\tilde{\Gamma}$

Ω_1

Γ

Figure 1.2.2   *Composite material domain and interface.*

1.2.4   *Initial, boundary and interface conditions in a practical example*

An example of a realistic physical problem is now used to illustrate how the correct boundary and initial conditions may be identified in practice. A tapered slab of aluminium bronze is cast in a resin bonded silica sand mould and a thermal analysis is to be made of the process. This problem involves the added complication that the metal is initially liquid and undergoes a change of phase during the transient. Methods of handling phase change problems will be introduced in Chapter 5 and the complications associated with the phase change process can be ignored for the purposes of this illustration. To reduce computational costs, the decision is made to undertake a two-dimensional analysis, and a vertical cross-section through the mould and casting, shown in Figure 1.2.3, is examined. In the actual problem, metal at 1156 °C is poured into the mould through the riser entrance CD. The simulation of the pouring stage can be expected to pose additional difficult problems and so the conduction analysis will start from the time when the mould is full, with the metal taken to be at a uniform initial temperature of 1156 °C. The external boundary EFGHAB of the sand will be assumed to be maintained at a constant temperature of 20 °C while the initial temperature in the sand is also taken to be at this value. The upper boundary, CD, of the riser will be held constant at a temperature 1156 °C, while the sides BC and DE of the riser will be assumed to be perfectly insulated and therefore subjected to a boundary condition of zero normal heat flux. The initial temperature at the interface between the mould and the metal is taken to be 1080 °C, which is the liquidus temperature of the metal and perfect conduction is assumed at the interface during the transient so that the boundary conditions of equation (1.2.20) may be applied.

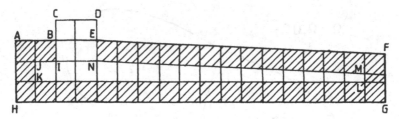

Figure 1.2.3 *Example of metal solidification. BIJKLMNE: interface between metal and mould.*

With the boundary, interface and initial conditions completely specified, the heat conduction equation can, in theory, be solved and the time variation of the temperature through the mould and the metal determined. The values of the thermal properties k and ρc of the metal and the mould will need to be specified before the solution process can begin.

1.3 VARIATIONAL FORMULATION

In the preceding section, we have outlined the classical formulation of the problem of heat conduction. This formulation is the starting point for certain approximate solution methods, such as the finite difference method, but is not appropriate for the finite element methods to be discussed in this text. For this reason, we will now replace the classical formulation by an equivalent variational formulation. It is this formulation which we shall employ as the starting point for the development of approximate solution techniques.

1.3.1 Steady problems

To illustrate the construction of a variational formulation, we begin by considering steady-state heat conduction in a domain, Ω, which consists of a single material. As in Figure 1.2.1 the temperature, T, will be assumed to be prescribed on the portion Γ_1 of the boundary curve, Γ, and the normal heat flux, q, will be prescribed on the remaining portion, Γ_2. The classical statement of this problem is then: determine $T(\mathbf{x})$ such that

$$\operatorname{div}(k \operatorname{grad} T) + Q = 0 \qquad \text{in } \Omega \qquad (1.3.1)$$

$$T = f(\mathbf{x}) \qquad \text{for } \mathbf{x} \text{ on } \Gamma_1 \qquad (1.3.2)$$

$$q(\mathbf{x}) = -(k \operatorname{grad} T)\cdot\mathbf{n} = -k\frac{\partial T}{\partial n} = \aleph(\mathbf{x}, T) \qquad \text{for } \mathbf{x} \text{ on } \Gamma_2 \qquad (1.3.3)$$

Here $k, f, \aleph$ and Q are prescribed functions. A variational formulation of the problem requires the introduction of a set, $\mathcal{T}$, of trial functions and a set, $\mathcal{W}$, of weighting functions. At this point, we will assume that $\mathcal{T}$ consists of all functions that satisfy the boundary conditions of (1.3.2) and (1.3.3) and a strong variational formulation of the above problem is then: find T in the trial function set $\mathcal{T}$ such that

$$\int_{\Omega} [\mathrm{div}(k \, \mathrm{grad} \, T) + Q] W \, d\Omega = 0 \qquad (1.3.4)$$

for each W in the weighting function set $\mathcal{W}$. An additional requirement now is that the functions which belong to the trial and weighting function sets must be sufficiently continuous so that the integral appearing in equation (1.3.4) is well defined. If we work with the largest possible sets, then the appropriate requirement here is that the weighting functions may be discontinuous but the trial functions must have continuous first derivatives. We use the notation that the weighting functions must show C^{-1} continuity and the trial functions must show C^1 continuity. The trial and weighting function sets can then be completely defined by

$$\mathcal{T} = \left\{ T \, | \, T = f(\mathbf{x}) \text{ on } \Gamma_1; q(\mathbf{x}) = -k\frac{\partial T}{\partial n} = \aleph(\mathbf{x}, T) \text{ on } \Gamma_2; T \text{ is } C^1 \right\} \quad (1.3.5)$$

$$\mathcal{W} = \{ W \, | \, W \text{ is } C^{-1} \} \qquad (1.3.6)$$

By the fundamental theorem of the variational calculus, the function $T(\mathbf{x})$ which satisfies the formulation of equation (1.3.4) is a solution of (1.3.1). The definition of the trial function set automatically ensures that this function also satisfies the conditions (1.3.2) and (1.3.3). Hence, the function $T(\mathbf{x})$ which satisfies this variational formulation is also the function which is the solution of the classical statement. For this variational formulation, the boundary conditions of equations (1.3.2) and (1.3.3) are termed essential, as they have to be satisfied by all members of the trial function set.

In the formulation presented above, we have determined the required solution from among the set of trial functions which satisfies the problem boundary conditions and have used the variational statement to impose the additional requirement of the satisfaction of the governing differential equation. However, alternative variational formulations of the problem, in which we widen the definition of the trial function set, are also possible. For example, relaxing completely the requirement that members of the trial function set should satisfy the problem boundary conditions, we can work with the trial and weighting function sets

$$\mathcal{T} = \{ T \, | \, T \text{ is } C^1 \} \qquad (1.3.7)$$

$$\mathcal{W} = \{ W \, | \, W \text{ is } C^{-1} \} \qquad (1.3.8)$$

and the variational formulation: find T in $\mathscr{T}$ such that

$$\int_{\Omega} [\operatorname{div}(k \operatorname{grad} T) + Q] W \, d\Omega + \Psi \int_{\Gamma_1} [T - f] W \, d\Gamma + \Phi \int_{\Gamma_2} \left[k \frac{\partial T}{\partial n} + \aleph \right] W \, d\Gamma = 0$$

(1.3.9)

for every W in $\mathscr{W}$. In this equation, Ψ and Φ are arbitrary constants. The fundamental theorem of the variational calculus can again be used to deduce that the function $T(\mathbf{x})$, which is the solution of this formulation, will satisfy equation (1.3.1) everywhere in Ω, the boundary condition of equation (1.3.2) everywhere on Γ_1, and the boundary condition of equation (1.3.3) everywhere on Γ_2. Hence this function is also the solution of the classical formulation. Here, the boundary conditions of (1.3.2) and (1.3.3) can be termed natural, as they do not have to be satisfied by the members of the trial function set but are imposed via the variational formulation.

An alternative, so-called weak, formulation of the problem, results from manipulating equation (1.3.9), assuming that the functions involved are sufficiently smooth. Using the result

$$[\operatorname{div}(k \operatorname{grad} T)] W = \operatorname{div}[W k \operatorname{grad} T] - k \operatorname{grad} T \cdot \operatorname{grad} W \qquad (1.3.10)$$

and the Divergence Theorem, we produce, from equation (1.3.9), the equation

$$\int_{\Gamma_1 + \Gamma_2} k \frac{\partial T}{\partial n} W \, d\Gamma - \int_{\Omega} [k \operatorname{grad} T \cdot \operatorname{grad} W - QW] \, d\Omega$$

$$+ \Psi \int_{\Gamma_1} [T - f] W \, d\Gamma + \Phi \int_{\Gamma_2} \left[k \frac{\partial T}{\partial n} + \aleph \right] W \, d\Gamma = 0 \qquad (1.3.11)$$

which can be simplified, if we adopt the value $\Phi = -1$, to

$$\int_{\Omega} k \operatorname{grad} T \cdot \operatorname{grad} W \, d\Omega = \int_{\Omega} QW \, d\Omega + \int_{\Gamma_1} W k \frac{\partial T}{\partial n} \, d\Gamma - \int_{\Gamma_2} W \aleph \, d\Gamma$$

$$+ \Psi \int_{\Gamma_1} [T - f] W \, d\Gamma \qquad (1.3.12)$$

A further simplification is possible if we also require that the members of the trial function set, $\mathscr{T}$, satisfy the boundary condition of equation (1.3.2). In this case, the variational formulation becomes: find T in $\mathscr{T}$ such that

$$\int_{\Omega} k \operatorname{grad} T \cdot \operatorname{grad} W \, d\Omega = \int_{\Omega} QW \, d\Omega + \int_{\Gamma_1} W k \frac{\partial T}{\partial n} \, d\Gamma - \int_{\Gamma_2} W \aleph \, d\Gamma \qquad (1.3.13)$$

for every W in the weighting function set $\mathscr{W}$. It is apparent that, for this variational formulation, the continuity requirements which are placed upon the members of the trial function set are reduced, whereas increased continuity requirements are placed upon members of the weighting function set. In fact, the requirement here is that members of both sets must be continuous (i.e. they must exhibit C^0 continuity) that the sets can be defined as

$$\mathcal{T} = \{T \mid T = f(\mathbf{x}) \text{ on } \Gamma_1; T \text{ is } C^0\} \tag{1.3.14}$$

$$\mathcal{W} = \{W \mid W \text{ is } C^0\} \tag{1.3.15}$$

A further simplification follows if the set of weighting functions is restricted to consist of those functions W which satisfy the homogeneous form of the boundary condition of equation (1.3.2), i.e.

$$W = 0 \quad \text{on } \Gamma_1 \tag{1.3.16}$$

Now, equation (1.3.13) implies that we look for T in $\mathcal{T}$ such that

$$\int_\Omega k \operatorname{grad} T \cdot \operatorname{grad} W \, d\Omega = \int_\Omega Q W \, d\Omega - \int_{\Gamma_2} W \aleph \, d\Gamma \tag{1.3.17}$$

for every W in $\mathcal{W}$. For this formulation, the appropriate trial and weighting function sets are

$$\mathcal{T} = \{T \mid T = f(\mathbf{x}) \text{ on } \Gamma_1; T \text{ is } C^0\} \tag{1.3.18}$$

$$\mathcal{W} = \{W \mid W = 0 \text{ on } \Gamma_1; W \text{ is } C^0\} \tag{1.3.19}$$

and it is apparent that there is now a close relationship between the two function sets. This close relationship is an attractive feature of this weak formulation for the heat conduction problem. This weak formulation is frequently preferred for both mathematical reasons and for practical applications.

For the weak formulation of equation (1.3.17)–(1.3.19), the boundary condition (1.3.2) is an essential boundary condition, as it is directly imposed upon the members of the trial function set. On the other hand, the boundary condition (1.3.3), which is automatically satisfied by the function $T(\mathbf{x})$ which is determined by the formulation, is a natural boundary condition as it is not directly imposed upon the members of the trial function set. The identification of any natural boundary conditions corresponding to a particular variational formulation can have important practical implications when an approximate solution of the problem is being attempted, as it can simplify considerably the construction of the trial function set.

1.3.2 Multi-material problems

For steady heat conduction in a region Ω consisting of two (or more) separate materials, it has been shown in equation (1.2.20) that the classical formulation requires the satisfaction of certain additional conditions at material interfaces. To consider the effect of this on our variational formulation, we assume that Ω is made up of two regions, Ω_1 and Ω_2, with material interface $\tilde{\Gamma}$, as shown in Figure 1.2.2. The classical formulation is then: determine a continuous function

$T(\mathbf{x})$ such that

$$\operatorname{div}(k \operatorname{grad} T) + Q = 0 \qquad \text{in } \Omega \qquad (1.3.20)$$

$$T = f(\mathbf{x}) \qquad \text{for } \mathbf{x} \text{ on } \Gamma_1 \qquad (1.3.21)$$

$$q(\mathbf{x}) = -(k \operatorname{grad} T) \cdot \mathbf{n} = -k \frac{\partial T}{\partial n} = \aleph(\mathbf{x}, T) \qquad \text{for } \mathbf{x} \text{ on } \Gamma_2 \qquad (1.3.22)$$

$$(k \operatorname{grad} T)_1 \cdot \tilde{\mathbf{n}} = (k \operatorname{grad} T)_2 \cdot \tilde{\mathbf{n}} \qquad \text{for } \mathbf{x} \text{ on } \tilde{\Gamma} \qquad (1.3.23)$$

where the subscripts 1 and 2 refer to conditions appropriate to the regions Ω_1 and Ω_2 respectively and where $k_1, k_2, f, \aleph$ and Q are prescribed functions. We will demonstrate that an appropriate weak formulation for this problem is again the formulation implied by (1.3.17)–(1.3.19). This can be proved by noting that, using equation (1.3.10),

$$\int_\Omega k \operatorname{grad} T \cdot \operatorname{grad} W \, d\Omega = \int_{\Omega_1 + \Omega_2} [\operatorname{div}(Wk \operatorname{grad} T) - W \operatorname{div}(k \operatorname{grad} T)] \, d\Omega$$

$$(1.3.24)$$

and it follows, using the Divergence Theorem over the regions Ω_1 and Ω_2 separately, that

$$\int_\Omega k \operatorname{grad} T \cdot \operatorname{grad} W \, d\Omega = -\int_\Omega W \operatorname{div}(k \operatorname{grad} T) \, d\Omega + \int_{\Gamma_2} Wk \operatorname{grad} T \cdot \mathbf{n} \, d\Gamma$$

$$+ \int_{\tilde{\Gamma}} W[|k \operatorname{grad} T \cdot \tilde{\mathbf{n}}|] \, d\Gamma \qquad (1.3.25)$$

where, for $\mathbf{x}$ on $\tilde{\Gamma}$, $[|f(\mathbf{x})|] = f_2(\mathbf{x}) - f_1(\mathbf{x})$ denotes the change in f across $\tilde{\Gamma}$. Thus the variational formulation of equations (1.3.17)–(1.3.19) is, in this case, equivalent to finding the function T in the trial function set $\mathcal{T}$ of equation (1.3.18) which is such that

$$-\int_\Omega [\operatorname{div}(k \operatorname{grad} T) + Q]W \, d\Omega + \int_{\Gamma_2} [k \operatorname{grad} T \cdot \mathbf{n} + \aleph]W \, d\Gamma$$

$$+ \int_{\tilde{\Gamma}} [|k \operatorname{grad} T \cdot \tilde{\mathbf{n}}|]W \, d\Gamma = 0 \qquad (1.3.26)$$

for all W in the weighting function set $\mathcal{W}$ of equation (1.3.19). This statement implies that the function $T(\mathbf{x})$ which satisfies the weak variational formulation will also be the solution of the classical problem defined by equations (1.3.20)–(1.3.23). The important feature to note is that both the boundary condition on Γ_2 and the interface condition on $\tilde{\Gamma}$ are natural boundary conditions for this weak formulation.

1.3.3 Transient problems

The inclusion of the effects of the time variable can be made directly within the variational formulations outlined above, provided that Ω and Γ now refer to the full space–time domain and its boundaries respectively. However, this is not the approach generally followed in this text, where the main intention is to treat the space and the time domains separately. In this case, the time t is considered as a real parameter and we develop a family of one-parameter variational formulations in t. We illustrate this approach by considering a problem, which is to be solved for a single material within a spatial domain Ω with spatial boundary Γ, whose classical formulation is: find $T(\mathbf{x}, t)$ such that

$$\operatorname{div}(k \operatorname{grad} T) + Q = \rho c \frac{\partial T}{\partial t} \qquad \text{for } \mathbf{x} \text{ in } \Omega, t > 0 \tag{1.3.27}$$

$$T = f(\mathbf{x}, t) \qquad \text{for } \mathbf{x} \text{ on } \Gamma_1, t > 0 \tag{1.3.28}$$

$$q(\mathbf{x}, t) = -(k \operatorname{grad} T) \cdot \mathbf{n} = -k \frac{\partial T}{\partial n} = \aleph(\mathbf{x}, T, t) \qquad \text{for } \mathbf{x} \text{ on } \Gamma_2, t > 0 \tag{1.3.29}$$

$$T = g(\mathbf{x}) \qquad \text{for } \mathbf{x} \text{ in } \Omega, t = 0 \tag{1.3.30}$$

Here $\rho, c, k, f, g, \aleph$ and Q are prescribed functions. To determine the solution over the interval $0 < t < \tau$, we define trial and weighting function sets by

$$\mathscr{T} = \{T(\mathbf{x}, t) \mid T = f(\mathbf{x}, t) \text{ on } \Gamma_1, 0 < t < \tau; \ T = g(\mathbf{x}) \text{ in } \Omega, t = 0; \ T \text{ is } C^0\} \tag{1.3.31}$$

$$\mathscr{W} = \{W(\mathbf{x}) \mid W = 0 \text{ on } \Gamma_1; \ W \text{ is } C^0\} \tag{1.3.32}$$

and we consider the variational formulation: find $T(\mathbf{x}, t)$ in $\mathscr{T}$ such that

$$\int_{\Omega} \left[\rho c \frac{\partial T}{\partial t} W + k \operatorname{grad} T \cdot \operatorname{grad} W \right] d\Omega = \int_{\Omega} Q W \, d\Omega - \int_{\Gamma_2} W \aleph \, d\Gamma \tag{1.3.33}$$

for every t in the range $0 < t < \tau$ and for every $W(\mathbf{x})$ in $\mathscr{W}$. By employing equation (1.3.24) and the Divergence Theorem on the left-hand side of this equation, it can be demonstrated that the solution T of this variational problem will also be the solution of the classical problem of equations (1.3.27)–(1.3.30). It should be noted that the boundary condition of equation (1.3.28) and the initial condition of equation (1.3.30) are essential conditions within this formulation, while the boundary condition of equation (1.3.29) is natural.

1.4 THE GALERKIN APPROXIMATE SOLUTION METHOD

The integral weak variational formulation of the heat conduction problem is the starting point for the development of the Galerkin method for constructing approximate solutions. We illustrate the approach by concentrating upon a particular example and we consider the solution of the steady linear heat conduction problem of equations (1.3.1)–(1.3.3), using initially the variational formulation of equation (1.3.17)–(1.3.19). We begin by determining any function $\psi(\mathbf{x})$ which satisfies the essential boundary condition on Γ_1 i.e.

$$\psi(\mathbf{x}) = f(\mathbf{x}) \qquad \text{for } \mathbf{x} \text{ on } \Gamma_1 \tag{1.4.1}$$

and choose a complete set of continuous approximating functions $N_1, N_2, \ldots,$ such that each member N_j of this set satisfies the homogeneous form of the boundary condition on Γ_1 i.e.

$$N_j(\mathbf{x}) = 0 \qquad \text{for } \mathbf{x} \text{ on } \Gamma_1 \tag{1.4.2}$$

The completeness requirement guarantees that these approximating functions form a basis for the trial and weighting function sets of (1.3.18) and (1.3.19), i.e.

$$\mathscr{T} = \left\{ T \,|\, T = \psi + \sum_{j=1}^{\infty} a_j N_j; \, T = f(\mathbf{x}) \text{ on } \Gamma_1; \, T \text{ is } C^0 \right\} \tag{1.4.3}$$

$$\mathscr{W} = \left\{ W \,|\, W = \sum_{j=1}^{\infty} b_j N_j; \, W = 0 \text{ on } \Gamma_1; \, W \text{ is } C^0 \right\} \tag{1.4.4}$$

where $a_1, a_2, \ldots,$ and $b_1, b_2, \ldots,$ are constants. To construct an approximate solution to the problem, we consider subsets $\mathscr{T}^{(M)}$ and $\mathscr{W}^{(M)}$, of finite dimension M, of the trial and weighting function sets, which are defined by

$$\mathscr{T}^{(M)} = \left\{ \hat{T} \,|\, \hat{T} = \psi + \sum_{j=1}^{M} a_j N_j; \, \hat{T} = f(\mathbf{x}) \text{ on } \Gamma_1; \, \hat{T} \text{ is } C^0 \right\} \tag{1.4.5}$$

$$\mathscr{W}^{(M)} = \left\{ W \,|\, W = \sum_{j=1}^{M} b_j N_j; \, W = 0 \text{ on } \Gamma_1; \, W \text{ is } C^0 \right\} \tag{1.4.6}$$

The Galerkin approximate solution is determined by employing the weak variational formulation of (1.3.17)–(1.3.19) as follows: find $\hat{T}$ in $\mathscr{T}^{(M)}$ such that

$$\int_{\Omega} k \, \text{grad} \, \hat{T} \cdot \text{grad} \, W \, d\Omega = \int_{\Omega} Q W \, d\Omega - \int_{\Gamma_2} W \aleph \, d\Gamma \tag{1.4.7}$$

for every W in the weighting function set $\mathscr{W}^{(M)}$. It can be observed from (1.4.6) that every W in $\mathscr{W}^{(M)}$ is formed as a linear combination of the approximating functions $N_1, N_2, \ldots, N_M$ and hence the requirement of equation (1.4.7) is equivalent to the requirement that the function $\hat{T}$ be such that

$$\int_{\Omega} k \, \text{grad} \, \hat{T} \cdot \text{grad} \, N_i \, d\Omega = \int_{\Omega} Q N_i \, d\Omega - \int_{\Gamma_2} N_i \aleph \, d\Gamma \qquad \text{for } i = 1, 2, \ldots, M \tag{1.4.8}$$

Inserting the assumed form for $\hat{T}$ from equation (1.4.5), it follows that

$$\int_{\Omega} k \operatorname{grad} \psi \cdot \operatorname{grad} N_i \, d\Omega + \sum_{j=1}^{M} \left\{ \int_{\Omega} k \operatorname{grad} N_j \cdot \operatorname{grad} N_i \, d\Omega \right\} a_j$$

$$= \int_{\Omega} Q N_i \, d\Omega - \int_{\Gamma_2} N_i \aleph \, d\Gamma \qquad \text{for } i = 1, 2, \dots, M \qquad (1.4.9)$$

This set of M equations can be expressed in the matrix form

$$\mathbf{Ka} = \mathbf{r} \qquad (1.4.10)$$

where $\mathbf{a}$ is the vector defined by

$$\mathbf{a}^{\mathrm{T}} = (a_1, a_2, \dots, a_M) \qquad (1.4.11)$$

and $\mathbf{r}$ is a vector with typical entry

$$r_i = \int_{\Omega} Q N_i \, d\Omega - \int_{\Gamma_2} N_i \aleph \, d\Gamma - \int_{\Omega} k \operatorname{grad} \psi \cdot \operatorname{grad} N_i \, d\Omega \qquad (1.4.12)$$

The $M \times M$ matrix $\mathbf{K}$ will, in general, be full with typical entry

$$K_{ij} = \int_{\Omega} k \operatorname{grad} N_j \cdot \operatorname{grad} N_i \, d\Omega \qquad (1.4.13)$$

The solution of the equation system (1.4.10) for the constants $a_1, a_2, \dots, a_M$ completes the determination of the Galerkin approximate solution $\hat{T}$ in the subset $\mathscr{T}^{(M)}$.

1.4.1 Optimality of the Galerkin approximation

For the problem of steady linear heat conduction, we are able to demonstrate, rather easily, that the Galerkin approximation possesses a certain optimality property. If

$$E = E(\mathbf{x}) = T - \hat{T} \qquad (1.4.14)$$

denotes the error in the Galerkin approximation and

$$E_S = E_S(\mathbf{x}) = T - \hat{S} \qquad (1.4.15)$$

is the error which results when any other function $\hat{S}$ in $\mathscr{T}^{(M)}$ is used as an approximation to T, then the optimality result can be expressed as

$$\| E \| \leqslant \| E_S \| \qquad (1.4.16)$$

where $\| \cdot \|$ denotes an appropriate measure, i.e. among the whole set of functions $\mathscr{T}^{(M)}$, the function constructed according to the Galerkin procedure is the best approximation possible according to the measure.

From equation (1.3.17), the exact solution T satisfies

$$\int_\Omega k\,\mathrm{grad}\,T\cdot\mathrm{grad}\,W\,d\Omega = \int_\Omega QW\,d\Omega - \int_{\Gamma_2} W\aleph\,d\Gamma \qquad (1.4.17)$$

for every W in the weighting function set $\mathscr{W}$. Since $\mathscr{W}^{(M)}$ is a subset of $\mathscr{W}$, this equation must also be valid for every W in $\mathscr{W}^{(M)}$. From equation (1.4.7), the Galerkin approximation $\hat{T}$ satisfies

$$\int_\Omega k\,\mathrm{grad}\,\hat{T}\cdot\mathrm{grad}\,W\,d\Omega = \int_\Omega QW\,d\Omega - \int_{\Gamma_2} W\aleph\,d\Gamma \qquad (1.4.18)$$

for every W in the weighting function set $\mathscr{W}^{(M)}$. Subtracting these two equations, it follows that

$$\int_\Omega k\,\mathrm{grad}\,E\cdot\mathrm{grad}\,W\,d\Omega = 0 \qquad (1.4.19)$$

for every $\overset{\bullet}{W}$ in the weighting function set $\mathscr{W}^{(M)}$. If we define a measure according to

$$\|f\|^2 = \int_\Omega k\,\mathrm{grad}\,f\cdot\mathrm{grad}\,f\,d\Omega \qquad (1.4.20)$$

the result

$$E_S = T - \hat{S} = (T - \hat{T}) + (\hat{T} - \hat{S}) = E + (\hat{T} - \hat{S}) \qquad (1.4.21)$$

leads to

$$\|E_S\|^2 = \int_\Omega k\,\mathrm{grad}\,E\cdot\mathrm{grad}\,E\,d\Omega + 2\int_\Omega k\,\mathrm{grad}\,E\cdot\mathrm{grad}(\hat{T} - \hat{S})\,d\Omega$$
$$+ \int_\Omega k\,\mathrm{grad}(\hat{T} - \hat{S})\cdot\mathrm{grad}(\hat{T} - \hat{S})\,d\Omega \qquad (1.4.22)$$

Since both $\hat{T}$ and $\hat{S}$ belong to $\mathscr{T}^{(M)}$, it can be observed, from equations (1.4.5) and (1.4.6), that $\hat{T} - \hat{S}$ belongs to $\mathscr{W}^{(M)}$. Hence, from equation (1.4.19), we deduce that

$$\int_\Omega k\,\mathrm{grad}\,E\cdot\mathrm{grad}(\hat{T} - \hat{S})\,d\Omega = 0 \qquad (1.4.23)$$

and it follows that equation (1.4.22) can be written as

$$\|E_S\|^2 = \|E\|^2 + \|\hat{T} - \hat{S}\|^2 \qquad (1.4.24)$$

The optimality result of equation (1.4.16) is now apparent since, from the definition of the measure in equation (1.4.20), $\|\hat{T} - \hat{S}\|^2$ is strictly non-negative.

1.4.2 Convergence of the Galerkin approximation

The Galerkin method, as outlined above, can be used with confidence in practice to construct approximate solutions to problems involving heat conduction if the convergence of the procedure can be demonstrated, i.e. if it can be proved that $\hat{T} \to T$ as the dimension M of the subsets $\mathcal{T}^{(M)}$ and $\mathcal{W}^{(M)}$ increases. The proof of convergence is not straightforward, but it is possible to show that, for linear heat conduction problems, convergence occurs according to the measure defined by equation (1.4.20), i.e. that $\| T - \hat{T} \| \to 0$ as $M \to \infty$.

1.5 FINITE ELEMENT APPROXIMATING FUNCTIONS IN ONE DIMENSION

It is apparent from the preceding discussion that the analyst who is interested in obtaining approximate solutions to heat conduction problems by the variational approach is immediately faced with the difficulty of choosing a suitable set of approximating functions. The finite element method helps the analyst in this respect by providing a systematic way of constructing a polynomial approximating function set of any desired dimension, M, by utilising the results of interpolation theory.

Consider the construction of an approximate solution to the one-dimensional steady linear heat conduction problem defined by

$$\frac{d}{d\xi}\left(k \frac{dT}{d\xi} \right) + Q = 0 \qquad -1 < \xi < 1 \tag{1.5.1}$$

$$T = f \qquad \text{at } \xi = -1 \tag{1.5.2}$$

$$q = -k\frac{dT}{dn} = -k\frac{dT}{d\xi} = \aleph \qquad \text{at } \xi = 1 \tag{1.5.3}$$

where k and Q are prescribed functions and f and $\aleph$ are prescribed constants. To obtain a solution by the finite element method, we begin by placing a set of $M + 1$ points, or nodes, at selected points $\xi_1, \xi_2, \ldots, \xi_{M+1}$ on the region, or element, defined by $-1 \leqslant \xi \leqslant 1$. One of these nodes is located at $\xi = -1$ and another is located at $\xi = 1$. For present purposes, the nodes are numbered according to the numbering system shown in Figure 1.5.1. With each node j we define the associated Lagrange interpolation polynomial $\mathcal{L}_j^{(M)}(\xi)$, of degree M, according to the requirements that

$$\mathcal{L}_j^{(M)}(\xi) = \begin{cases} 1 & \text{if } \xi = \xi_j \\ 0 & \text{if } \xi = \xi_k, k \neq j \end{cases} \tag{1.5.4}$$

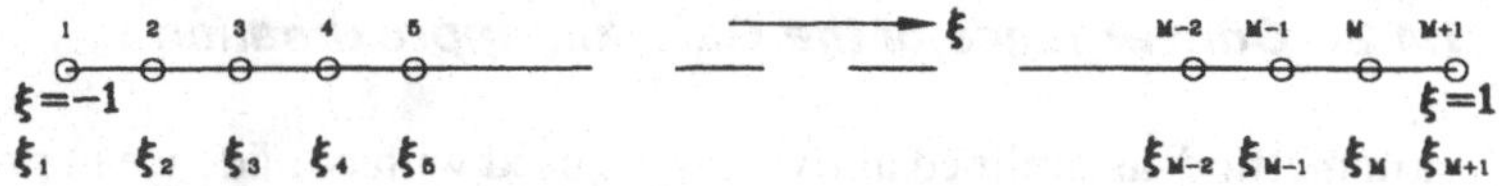

Figure 1.5.1   *Node numbering for one-dimensional problem using linear elements.*

which leads to the expression

$$\mathscr{L}_j^{(M)}(\xi) = \prod_{\substack{l=1 \\ l \neq j}}^{M+1} \frac{(\xi - \xi_l)}{(\xi_j - \xi_l)} \tag{1.5.5}$$

If we now adopt the choice $N_j = \mathscr{L}_j^{(M)}$, then the function

$$\hat{T} = \sum_{j=1}^{M+1} a_j N_j \tag{1.5.6}$$

is such that $\hat{T} = a_j$ when $\xi = \xi_j$, i.e. the coefficient a_j is the value of the function $\hat{T}$ at the point $\xi = \xi_j$. To denote this, we replace the representation of equation (1.5.6) by

$$\hat{T} = \sum_{j=1}^{M+1} T_j N_j \tag{1.5.7}$$

and adopt the definitions

$$\mathscr{T}^{(M)} = \left\{ \hat{T} \mid \hat{T} = \psi + \sum_{j=2}^{(M+1)} T_j N_j; \psi = f N_1; \hat{T} = f \text{ at } \xi = -1 \right\} \tag{1.5.8}$$

for the trial function set and

$$\mathscr{W}^{(M)} = \left\{ W \mid W = \sum_{j=2}^{(M+1)} b_j N_j; W = 0 \text{ at } \xi = -1 \right\} \tag{1.5.9}$$

for the weighting function set. Following the procedure outlined in Section 1.4, these finite dimensional trial and weighting functions sets can be used to determine an approximate solution to the problem of equations (1.5.1)–(1.5.3) by the Galerkin method. The resulting solution is the Galerkin finite element approximation which is obtained when a polynomial of order (or degree) M is employed over a single $M + 1$ noded finite element.

In the current context, the function N_j is termed the finite element shape function associated with node j and it is informative at this stage to examine some low order one-dimensional elements in greater detail.

1.5.1   The linear element

The linear element, for which $M = 1$, has two nodes, located at $\xi = -1$ and at $\xi = 1$, as shown in Figure 1.5.2. The corresponding shape functions are linear

Figure 1.5.2   *One-dimensional linear element.*

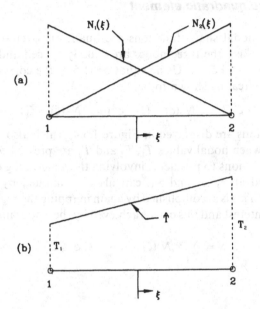

Figure 1.5.3   *Shape functions and form of function for one-dimensional linear element.*

over the element and equation (1.5.5) can be used to construct expressions for these functions as

$$N_1(\xi) = (1 - \xi)/2 \qquad N_2(\xi) = (1 + \xi)/2 \qquad (1.5.10)$$

These shape functions are shown in Figure 1.5.3, which also shows the form of the function $\hat{T}$ when nodal values T_1 and T_2 are prescribed.

The problem of equations (1.5.1)–(1.5.3) is defined over the region $-1 \leqslant \xi \leqslant 1$. Problems involving the more general region $x_1 \leqslant x \leqslant x_2$ can be handled by mapping the region $-1 \leqslant \xi \leqslant 1$ into the region of interest. An isoparametric mapping accomplishes this by employing the element shape functions and defining the mapping by

$$x = \sum_{j=1}^{2} x_j N_j(\xi) \qquad -1 \leqslant \xi \leqslant 1 \qquad (1.5.11)$$

In this case, the trial and weighting function sets of (1.5.8) and (1.5.9) will become

$$\mathcal{T}^{(1)} = \{\hat{T}(x) | \hat{T}(x) = \psi + T_2 N_2(\xi); \psi = f N_1(\xi); \hat{T} = f \text{ at } x = x_1\} \quad (1.5.12)$$

$$\mathcal{W}^{(1)} = \{W(x) | W(x) = b_2 N_2(\xi); W = 0 \text{ at } x = x_1\} \qquad (1.5.13)$$

with the relation between the coordinates x and ξ being given by the mapping of equation (1.5.11).

1.5.2 The quadratic element

When $M = 2$, the nodal shape functions are quadratic over the element. We consider the case where the three nodes are equally spaced and located at the points $\xi = -1$, $\xi = 0$ and $\xi = 1$. Using equation (1.5.5), the corresponding nodal shape functions are readily shown to be

$$N_1(\xi) = -\xi(1 - \xi)/2 \qquad N_2(\xi) = (1 - \xi^2) \qquad N_3(\xi) = \xi(1 + \xi)/2 \qquad (1.5.14)$$

These shape functions are displayed in Figure 1.5.4, which also shows the form of the function $\hat{T}$ when nodal values T_1, T_2 and T_3 are prescribed.

Approximate solutions to problems involving the general region $x_1 \leqslant x \leqslant x_3$, with nodes located at x_1, x_2 and x_3, can also be handled by using a single quadratic element. This is accomplished by again mapping the region $-1 \leqslant \xi \leqslant 1$ into the region of interest and this can be achieved by the isoparametric mapping

$$x = \sum_{j=1}^{3} x_j N_j(\xi) \qquad -1 \leqslant \xi \leqslant 1 \qquad (1.5.15)$$

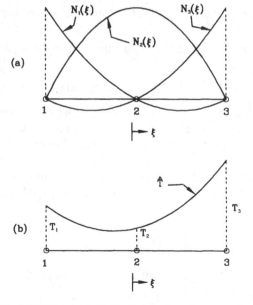

Figure 1.5.4   *Shape functions and form of function for one-dimensional quadratic element.*

The trial and weighting function sets of (1.5.8) and (1.5.9) now become

$$\mathscr{T}^{(2)} = \left\{ \hat{T}(x) | \, \hat{T}(x) = \psi + \sum_{j=2}^{3} T_j N_j(\xi); \psi = f N_1(\xi); \, \hat{T} = f \text{ at } x = x_1 \right\} \quad (1.5.16)$$

$$\mathscr{W}^{(2)} = \left\{ W(x) | \, W(x) = \sum_{j=2}^{3} b_j N_j(\xi); \, W = 0 \text{ at } x = x_{\Gamma} \right\} \quad (1.5.17)$$

with the relation between the coordinates x and ξ being given by the mapping of equation (1.5.15).

1.6 FINITE ELEMENT APPROXIMATING FUNCTIONS IN TWO DIMENSIONS

Now consider the construction of an approximate solution to a two-dimensional heat conduction problem on a square region defined by

$$\frac{\partial}{\partial \xi}\left(k \frac{\partial T}{\partial \xi}\right) + \frac{\partial}{\partial \eta}\left(k \frac{\partial T}{\partial \eta}\right) + Q = 0 \qquad -1 < \xi, n < 1 \quad (1.6.1)$$

$$T = f(\xi) \qquad \text{on } \eta = -1 \quad (1.6.2)$$

$$q = -k \frac{\partial T}{\partial n} = \aleph \qquad \text{on } \xi = \pm 1 \text{ and on } \eta = 1 \quad (1.6.3)$$

Here $k, f, \aleph$ and Q are prescribed functions. If an approximate solution to this problem is to be sought by applying the Galerkin procedure, a suitable polynomial approximating function set can be obtained by direct extension of the one-dimensional finite element ideas introduced previously. We select a set of $M + 1$ points $\xi_1, \xi_2, \ldots, \xi_{M+1}$, such that $\xi_1 = -1$ and $\xi_{M+1} = 1$, and a set of $M + 1$ points $\eta_1, \eta_2, \ldots, \eta_{M+1}$, such that $\eta_1 = -1$ and $\eta_{M+1} = 1$. Nodes, numbered so that the first $4M$ lie on the boundary and the first $M + 1$ lie on the line $\eta = -1$, as shown in Figure 1.6.1, are located at the points (ξ_k, η_l) and the associated Lagrange interpolation polynomials $\mathscr{L}_k^{(M)}(\xi)$ and $\mathscr{L}_l^{(M)}(\eta)$ are constructed. For node j, located at (ξ_k, η_l), the finite element shape function is defined by

$$N_j(\xi, \eta) = \mathscr{L}_k^{(M)}(\xi) \mathscr{L}_n^{(M)}(\eta) \quad (1.6.4)$$

It is readily demonstrated that this function has the value unity at node j and the value zero at all the other nodes. The trial function set for these approximating functions becomes

$$\mathscr{T}^{(M(M+1))} = \left\{ \hat{T} | \, \hat{T} = \psi + \sum_{j=M+2}^{(M+1)^2} T_j N_j; \psi = \sum_{j=1}^{M+1} f_j N_j; \, \hat{T} = \hat{f} \text{ on } \eta = -1 \right\} \quad (1.6.5)$$

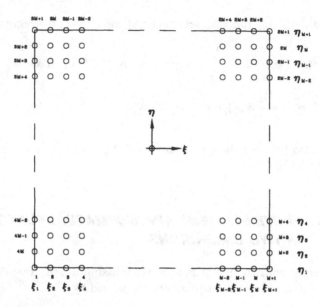

Figure 1.6.1   *Node numbering for two-dimensional problem using linear elements.*

and the corresponding weighting function set is

$$\mathscr{W}^{(M(M+1))} = \left\{ W \mid W = \sum_{j=M+2}^{(M+1)^2} b_j N_j; W = 0 \text{ on } \eta = -1 \right\} \tag{1.6.6}$$

Note now that the members $\hat{T}$ of the trial function set satisfy $\hat{T} = \hat{f}$ on the boundary $\eta = -1$, where

$$\hat{f}(\xi) = \sum_{j=1}^{M+1} f_j N_j(\xi, -1) \tag{1.6.7}$$

In this representation, f_j denotes the value of $f(\xi)$ at node j, i.e. at $\xi = \xi_j$, and (1.6.7) is the Lagrange interpolating polynomial of degree M which passes through the prescribed values $f_1, f_2, \ldots, f_{M+1}$ at the points $\xi_1, \xi_2, \ldots, \xi_{M+1}$.

It is informative at this stage to derive the shape functions and the corresponding trial and weighting function sets for some low order two-dimensional elements.

1.6.1   The linear element

The choice $M = 1$ leads to a four-noded square element with the nodes located at the corners, as shown in Figure 1.6.2. Equations (1.5.5) and (1.6.4) can be used to show that the corresponding nodal shape functions are

$$\begin{aligned} N_1 &= (1 - \xi)(1 - \eta)/4 & N_2 &= (1 + \xi)(1 - \eta)/4 \\ N_3 &= (1 + \xi)(1 + \eta)/4 & N_4 &= (1 - \xi)(1 + \eta)/4 \end{aligned} \tag{1.6.8}$$

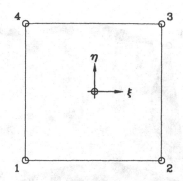

Figure 1.6.2 *Four-noded square element.*

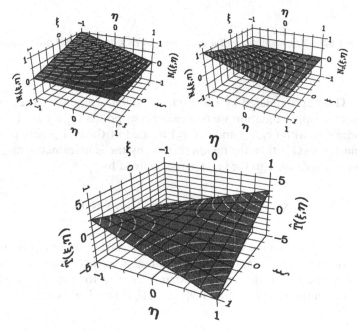

Figure 1.6.3 *Form of shape functions for four-noded element.*

The form of these shape functions is illustrated in Figure 1.6.3, which also shows the form of the function $\hat{T}$ when nodal values T_1, T_2, T_3 and T_4 are prescribed.

For the problem

$$\frac{\partial}{\partial x}\left(k\frac{\partial T}{\partial x}\right) + \frac{\partial}{\partial y}\left(k\frac{\partial T}{\partial y}\right) + Q = 0 \qquad \mathbf{x} \text{ in } \Omega \tag{1.6.9}$$

$$T = f(\mathbf{x}) \qquad \mathbf{x} \text{ on } \Gamma_1 \tag{1.6.10}$$

$$q = -k\frac{\partial T}{\partial n} = \aleph(\mathbf{x}) \qquad \mathbf{x} \text{ on } \Gamma_2 \tag{1.6.11}$$

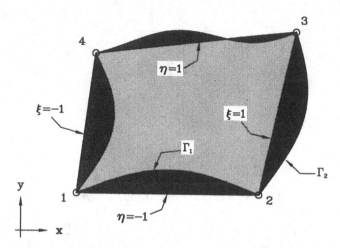

Figure 1.6.4 *Approximation of a given region by a four-noded quadrilateral.*

the region Ω is approximated by a single four-noded, straight-sided element. To accomplish the approximation we first select four points $(x_j, y_j); j = 1, 2, 3, 4$, on the boundary Γ, with (x_1, y_1) and (x_2, y_2) located at the end points of Γ_1. The approximation to Ω is then the region formed by the isoparametric mapping of the region $-1 \leqslant \xi, \eta \leqslant 1$ to the (x, y) plane, defined by

$$x = \sum_{j=1}^{4} x_j N_j(\xi, \eta) \qquad y = \sum_{j=1}^{4} y_j N_j(\xi, \eta) \qquad (1.6.12)$$

This process is illustrated in Figure 1.6.4, with the darkly shaded area indicating the error which is made in approximating a region in this manner.

When a single linear element is used in this way to construct a Galerkin approximate solution to problem (1.6.9)–(1.6.11), the trial and weighting function sets are

$$\mathscr{T}^{(2)} = \left\{ \hat{T}(x, y) | \, \hat{T}(x, y) = \psi + \sum_{j=3}^{4} T_j N_j(\xi, \eta); \psi = \sum_{j=1}^{2} f_j N_j(\xi, \eta); \hat{T} = \hat{f} \text{ on } \eta = -1 \right\}$$

$$(1.6.13)$$

$$\mathscr{W}^{(2)} = \left\{ W(x, y) | \, W(x, y) = \sum_{j=3}^{4} b_j N_j(\xi, \eta); W = 0 \text{ on } \eta = -1 \right\} \quad (1.6.14)$$

with the relation between the coordinates (x, y) and (ξ, η) being defined by the mapping of equation (1.6.12). In this way, the Galerkin solution is constructed over the approximated region in the (x, y) plane.

1.6.2 The quadratic element

We consider the choice $M = 2$, with the nodes located at the corners of the element, at the mid-points of the element sides and at the element centre. The nine nodes are numbered as in Figure 1.6.5, and the corresponding nodal shape functions follow from equations (1.5.5) and (1.6.4) as

$$N_1 = (\xi^2 - \xi)(\eta^2 - \eta)/4 \qquad N_2 = (1 - \xi^2)(\eta^2 - \eta)/2$$
$$N_3 = (\xi^2 + \xi)(\eta^2 - \eta)/4 \qquad N_4 = (\xi^2 + \xi)(1 - \eta^2)/2$$
$$N_5 = (\xi^2 + \xi)(\eta^2 + \eta)/4 \qquad N_6 = (1 - \xi^2)(\eta^2 + \eta)/2 \qquad (1.6.15)$$
$$N_7 = (\xi^2 - \xi)(\eta^2 + \eta)/4 \qquad N_8 = (\xi^2 - \xi)(1 - \eta^2)/2$$
$$N_9 = (1 - \xi^2)(1 - \eta^2)$$

The form of these shape functions is illustrated in Figure 1.6.6, which also shows the form of the function $\hat{T}$ when nodal values $T_1, T_2, \ldots, T_9$ are prescribed.

An approximate solution to the problem of equations (1.6.9)–(1.6.11) may be constructed by first employing an isoparametric mapping of a single nine-noded element to approximate the region of interest, Ω, in the (x, y) plane. This is accomplished by selecting eight points (x_j, y_j); $j = 1, 2, \ldots, 8$, on the boundary Γ, with (x_1, y_1) and (x_3, y_3) located at the end points of Γ_1 and (x_2, y_2) also located on Γ_1. A point (x_9, y_9) is also selected on the interior of Ω. The isoparametric relation

$$x = \sum_{j=1}^{9} x_j N_j(\xi, \eta) \qquad y = \sum_{j=1}^{9} y_j N_j(\xi, \eta) \qquad (1.6.16)$$

is then used to map the region $-1 \leqslant \xi, \eta \leqslant 1$ into the (x, y) plane. This process is illustrated in Figure 1.6.7, with the darkly shaded area indicating the error which is made in approximating the region in this manner.

When a single quadratic element is used in this way to construct a Galerkin approximate solution to problem (1.6.9)–(1.6.11), the trial and weighting function

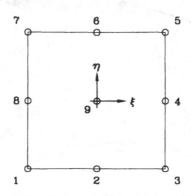

Figure 1.6.5   *Nine-noded quadrilateral element.*

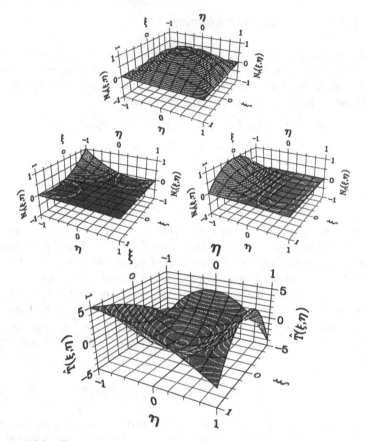

Figure 1.6.6 *Form of shape functions for nine-noded quadrilateral element.*

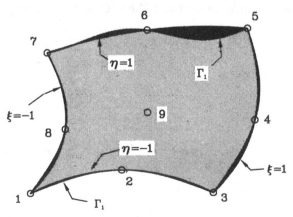

Figure 1.6.7 *Approximation of a given region by a nine-noded quadrilateral.*

sets are

$$\mathcal{T}^{(6)} = \left\{ \hat{T}(x,y) | \hat{T}(x,y) = \psi + \sum_{j=4}^{9} T_j N_j(\xi,\eta); \psi = \sum_{j=1}^{3} f_j N_j(\xi,\eta); \hat{T} = \hat{f} \text{ on } \eta = -1 \right\}$$

(1.6.17)

$$\mathcal{W}^{(6)} = \left\{ W(x,y) | W(x,y) = \sum_{j=4}^{9} b_j N_j(\xi,\eta); W = 0 \text{ on } \eta = -1 \right\} \quad (1.6.18)$$

with the relation between the coordinates (x,y) and (ξ,η) being defined by the mapping of equation (1.6.16).

1.6.3 The quadratic serendipity element

The serendipity family of elements provides an alternative approach to the problem of defining polynomial approximating functions over a square region. These elements possess a reduced number of internal nodes and the shape function associated with node j is determined directly from the requirement that

$$N_j = \begin{cases} 1 & \text{at node } j \\ 0 & \text{at node } k; k \neq j \end{cases}$$

(1.6.19)

We illustrate this approach for an eight-noded element, with the nodes located as shown in Figure 1.6.8. In this case, it is easy to confirm that the corresponding nodal shape functions are

$$\begin{aligned} N_1 &= (1-\xi)(1-\eta)(-1-\xi-\eta)/4 & N_2 &= (1-\xi^2)(1-\eta)/2 \\ N_3 &= (1+\xi)(1-\eta)(-1+\xi-\eta)/4 & N_4 &= (1+\xi)(1-\eta^2)/2 \\ N_5 &= (1+\xi)(1+\eta)(-1+\xi+\eta)/4 & N_6 &= (1-\xi^2)(1+\eta)/2 \\ N_7 &= (1-\xi)(1+\eta)(-1-\xi-\eta)/4 & N_8 &= (1-\xi)(1-\eta^2)/2 \end{aligned}$$

(1.6.20)

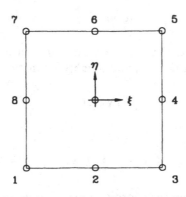

Figure 1.6.8 *Eight-noded quadratic serendipity element.*

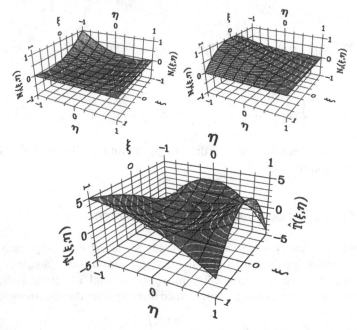

Figure 1.6.9 *Form of shape functions and function for eight-noded quadrilateral.*

The form of these shape functions is illustrated in Figure 1.6.9, which also shows the form of the function $\hat{T}$ when nodal values $T_1, T_2, \ldots, T_8$ are prescribed.

A single eight-noded element can be used to obtain an approximate solution to the problem of equations (1.6.9)–(1.6.11). By selecting eight points (x_j, y_j); $j = 1, 2, \ldots, 8$, on the boundary Γ, with (x_1, y_1) and (x_3, y_3) located at the end points of Γ_1 and (x_2, y_2) also located on Γ_1, an element can be created which approximates the region Ω by mapping the region $-1 \leqslant \xi, \eta \leqslant 1$ by the iso-parametric relation

$$x = \sum_{j=1}^{8} x_j N_j(\xi, \eta) \qquad y = \sum_{j=1}^{8} y_j N_j(\xi, \eta) \qquad (1.6.21)$$

When this element is used to construct a Galerkin approximate solution to problem (1.6.9)–(1.6.11), the trial and weighting function sets are

$$\mathcal{T}^{(5)} = \left\{ \hat{T}(x, y) \mid \hat{T}(x, y) = \psi + \sum_{j=4}^{8} T_j N_j(\xi, \eta); \psi = \sum_{j=1}^{3} f_j N_j(\xi, \eta); \hat{T} = \hat{f} \text{ on } \eta = -1 \right\}$$

$$(1.6.22)$$

$$\mathcal{W}^{(5)} = \left\{ W(x, y) \mid W(x, y) = \sum_{j=4}^{8} b_j N_j(\xi, \eta); W = 0 \text{ on } \eta = -1 \right\} \quad (1.6.23)$$

with the relation between the coordinates (x, y) and (ξ, η) now being defined by the mapping of equation (1.6.21).

1.7 A PRACTICAL IMPLEMENTATION OF THE GALERKIN FINITE ELEMENT METHOD

In the description of the Galerkin finite element method which has been presented above, it can be observed that the nodes on Γ_1, at which essential boundary conditions have to be imposed, are treated from the outset in a different fashion to the other nodes. This observation leads to unnecessary complexities when attempting to develop a computational implementation of the procedure.

To illustrate how these complexities can be overcome, consider the problem of solving the differential equation (1.6.1) using the element e shown in Figure 1.6.1. We begin by assuming that neither of the boundary conditions will be treated as essential conditions and work with the trial function set

$$\mathcal{T}^{((M+1)^2)} = \left\{ \hat{T} \,\middle|\, \hat{T} = \sum_{j=1}^{(M+1)^2} T_j N_j \right\} \tag{1.7.1}$$

and the weighting function set

$$\mathcal{W}^{((M+1)^2)} = \left\{ W \,\middle|\, W = \sum_{j=1}^{(M+1)^2} b_j N_j \right\} \tag{1.7.2}$$

An appropriate starting point for the construction of a finite element approximate solution is then to employ equation (1.3.12) as the basis of a variational formulation in the form: find $\hat{T}$ in $\mathcal{T}^{(M+1)^2}$ such that

$$\int_\Omega k \operatorname{grad} \hat{T} \cdot \operatorname{grad} N_i \, d\Omega = \int_\Omega Q N_i \, d\Omega - \int_{\Gamma_1} N_i \hat{q} \, d\Gamma - \int_{\Gamma_2} N_i \aleph \, d\Gamma$$

$$+ \Psi \int_{\Gamma_1} [\hat{T} - \hat{f}] N_i \, d\Gamma \tag{1.7.3}$$

where $\hat{q}$ is an approximation to the unknown value of the flux on Γ_1. Inserting the assumed form for $\hat{T}$ from equation (1.7.1), it follows that the finite element approximation is determined by the solution of the $(M+1) \times (M+1)$ matrix equation system

$$\mathbf{K}_e \mathbf{T}_e = \mathbf{r}_e - \mathbf{q}_e \tag{1.7.4}$$

where the vectors $\mathbf{T}_e$ and $\mathbf{q}_e$ are defined by

$$\mathbf{T}_e^T = (T_1, T_2, \ldots, T_{(M+1)^2}) \qquad \mathbf{q}_e^T = (q_1, q_2, \ldots, q_{4M}, 0, \ldots, 0) \tag{1.7.5}$$

and where

$$q_i = \int_\Gamma \hat{q} N_i \, d\Gamma \tag{1.7.6}$$

where $\hat{q}$ is now to be regarded as equal to the unknown heat flux on Γ_1 and equal to the specified heat flux (i.e. $\hat{q} = \aleph$) on Γ_2. The matrix $\mathbf{K}_e$ and the vector $\mathbf{r}_e$ have typical entries

$$K_{ij} = \int_\Omega k \operatorname{grad} N_j \cdot \operatorname{grad} N_i \, d\Omega$$

$$r_i = \int_\Omega Q N_i \, d\Omega + \Psi \int_{\Gamma_1} [\hat{T} - \hat{f}] N_i \, d\Gamma \qquad (1.7.7)$$

At this stage, we satisfy the boundary condition on Γ_1 by imposing the values

$$T_i = f_i \qquad \text{for } i = 1, 2, \ldots, M + 1 \qquad (1.7.8)$$

so that the integral over Γ_1 in equation (1.7.7) disappears from the formulation. Equation system (1.7.4) can then be solved for the approximations to the values of the temperature at the $M(M + 1)$ nodes not subject to the specified temperature boundary condition. It is readily demonstrated that these will be exactly the values determined by the direct application of the Galerkin method as outlined previously. At the same time, the solution process will give values for the approximation to the integrated flux, q_i, at all nodes at which the value of the temperature has been specified.

It can be observed that, if Q is represented by its Lagrange interpolant over Ω and if $\aleph$ is represented by its Lagrange interpolant on Γ_2, the approximations

$$\int_\Omega Q N_i \, d\Omega \approx \sum_{j=1}^{(M+1)^2} \left[\int_e N_j N_i \, d\Omega \right] Q_j \qquad \text{for } i = 1, 2, \ldots, (M+1)^2 \quad (1.7.9)$$

and

$$\int_{\Gamma_2} \aleph N_i \, d\Omega \approx \sum_{j=M+1}^{4M} \left[\int_{\Gamma_2} N_j N_i \, d\Gamma \right] \aleph_j \qquad \text{for } i = M+1, M+2, \ldots, 4M$$

$$(1.7.10)$$

can be employed to simplify the evaluation of the components of the vectors **r** defined in equation (1.7.7) and **q** in equation (1.7.6).

1.8 FURTHER OBSERVATIONS

We have now laid some of the important foundations related to the use of the finite element method for the approximate solution of heat conduction problems and we are ready to begin studying, in detail, the practical implementation of the method. However, before we embark on this study, a few further observations can be made on the theory which has been introduced in this chapter.

The reader will have observed that the shape functions for the two-dimensional linear and quadratic elements, introduced in the last section, contain additional terms to those that would be present in a complete linear or quadratic polynomial in ξ and η. Thus, for example, the linear element shape functions contain the bilinear term $\xi\eta$, while the quadratic Lagrangian element shape functions contain the terms $\xi^2\eta, \xi\eta^2$ and $\xi^2\eta^2$. The quadratic serendipity element contains the additional terms $\xi^2\eta$ and $\xi\eta^2$. These parasitic polynomial terms cannot be removed from the nodal shape functions for square elements. The construction of elements which are triangular in shape generally leads to nodal shape functions which are in the form of complete polynomials, but such elements will not be treated in any detail in this book.

It will also be apparent that the evaluation of the entries in the matrix $\mathbf{K}_e$ and the vector $\mathbf{r}_e$ of equation (1.7.7) will require the evaluation of integrals, over the approximated area of interest, when the solution of a problem such as (1.6.9)–(1.6.11) is attempted by the Galerkin method. We will be following an approach in which such integrals are evaluated approximately, by standard numerical integration techniques in the mapped (ξ, η) plane. This topic will be addressed in the next chapter.

The analyst who is interested in improving the accuracy of the computation can achieve this by increasing the dimension M of the trial function set, i.e. by employing a higher-order element. In practice, however, the approach which is normally adopted is to use low-order elements (such as the linear or quadratic elements introduced above) and to improve the solution quality by sub-dividing the region of interest into a number of elements, as illustrated for example in Figure 1.2.3, and to employ a piecewise (local) low-order interpolation of the solution over each element. The resulting approximate solution $\hat{T}$ will possess C^0 continuity, e.g. considering an assembly of quadratic serendipity elements, inter-element continuity of the approximation is assured as the three nodes located along any element side determine a unique quadratic expansion. This sub-division of the region of interest means that the process of evaluating integrals, such as those appearing in equation (1.7.7), will require further attention. The way in which this complication can be handled will be considered in the next chapter.

ADDITIONAL READING

Becker E. B., Carey G. F. and Oden J. T., *Finite Elements — An Introduction*, Volume 1, Prentice-Hall (1981).

Carslaw H. S. and Jaeger J. C., *Conduction of Heat in Solids* (2nd Edition), Oxford University Press (1959).

Johnson C., *Numerical Solution of Partial Differential Equations by the Finite Element Method*, Cambridge University Press (1987).

Kreyszig E., *Advanced Engineering Mathematics, 7th Edition*, John Wiley & Sons (1993).

Luikov A. V., *Analytical Heat Diffusion Theory*, Academic Press (1968).

Milne R. D., *Applied Functional Analysis — An Introductory Treatment*, Pitman Publishing (1980).

Prenter P. M., *Splines and Variational Methods*, Wiley (1975).

Reddy J. N., *An Introduction to the Finite Element Method*, McGraw-Hill (1984).

Strang G. and Fix G., *An Analysis of the Finite Element Method*, Prentice-Hall (1973).

Zienkiewicz O. C. and Morgan K., *Finite Elements and Approximation*, John Wiley & Sons (1983).

Zienkiewicz O. C. and Taylor R. L., *The Finite Element Method, 4th Edition*, Volume 1, McGraw-Hill (1989).

2 Linear Steady-state Problems

2.1 INTRODUCTION

In this chapter, we will show how problems involving linear steady-state heat conduction may be solved approximately by using simple finite elements and the Galerkin form of the weak variational formulation introduced in the previous chapter. Initially, problems will be solved by employing a single element and the approach will then be extended to deal with problems in which an assembly of elements, with piecewise polynomial interpolation of the temperature field, is employed.

We begin by focusing attention on problems involving one space dimension, in an attempt to develop concepts and to aid the understanding of the element assembly process. This assembly process, in which entries in the finite element equation system are evaluated by summing contributions from individual elements, is fundamental to the general finite element method. It will be seen that the element contributions may be defined generally in a straightforward manner, thus allowing the solution process to be readily programmed for computer use.

The stage is then set for the development of the method for problems involving more than one space dimension and the extension of the method to the solution of two-dimensional problems is presented. The further extension to three-dimensions is not considered here as it causes no new conceptual difficulties.

The element contributions are generally expressed as integrals of appropriate quantities over the element or its boundaries. For some of the examples considered in this chapter, it is possible to evaluate such integrals analytically. However, even when exact evaluation is possible, it is frequently more convenient, especially when employing the computer, to use numerical integration techniques. Therefore, this chapter also contains a description of how Gauss–Legendre numerical integration methods can be used to evaluate the element contributions.

2.2   A ONE-DIMENSIONAL PROBLEM SOLVED USING A SINGLE ELEMENT

We begin by demonstrating how the solution of a problem of steady heat conduction in one dimension can be obtained by using a single finite element. The problem consists of determining the temperature distribution resulting from steady-state heat conduction in a thin rod, of length L and thermal conductivity k, whose curved surface is thermally insulated (see Figure 2.2.1). The x axis is aligned with the longitudinal axis of the cylinder, with the origin at one end as shown. If heat is generated within the rod at a prescribed rate Q per unit length, then, as there is no heat loss across the curved surface, the temperature distribution through the rod is governed by the solution of the one-dimensional heat conduction equation

$$\frac{d}{dx}\left(k\frac{dT}{dx}\right) + Q = 0 \qquad 0 < x < L \tag{2.2.1}$$

General boundary conditions,

$$T = f \qquad \text{on } \Gamma_1 \tag{2.2.2}$$

$$q = \aleph \qquad \text{on } \Gamma_2 \tag{2.2.3}$$

will be assumed at this stage, where Γ_1 and Γ_2 denote either of the boundary points $x = 0$ or $x = L$. In addition, f is a prescribed constant and $\aleph$ may be a prescribed constant or a prescribed function of T, as in equations (1.2.18) or (1.2.19). The application of particular boundary conditions can be achieved by appropriate definition of Γ_1 and Γ_2 and the specification of values for f and $\aleph$.

2.2.1   The linear element

Initially, as in Section 1.5.1, we employ one linear element, e, to represent the region of interest. The nodes of this element are numbered 1, 2, as shown in Figure 2.2.2 and the element length $h_e = L$. The standard two-noded linear element, defined over the range $-1 \leqslant \xi \leqslant 1$, can be mapped into the element

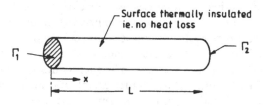

Figure 2.2.1   *Heat conduction in a thin rod.*

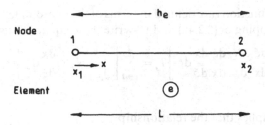

Figure 2.2.2 *Node numbers for a linear element.*

e, by means of the isoparametric mapping of equation (1.5.11), according to

$$x = \sum_{j=1}^{2} x_j N_j(\xi) = (x_1 + x_2)/2 + h_e\xi/2 = h_e/2 + h_e\xi/2 \qquad (2.2.4a)$$

since $x_1 = 0$ and $x_2 = h_e$ in this case. Here we have made use of the fact that the linear element shape functions, defined in equation (1.5.10), are

$$N_1(\xi) = (1 - \xi)/2 \qquad N_2(\xi) = (1 + \xi)/2 \qquad -1 \leqslant \xi \leqslant 1 \qquad (2.2.4b)$$

Following the procedure outlined in Section 1.7, we work with trial functions

$$\hat{T} = \sum_{j=1}^{2} T_j N_j(\xi) \qquad (2.2.5)$$

and the finite element approximation is determined from the requirement that

$$\int_e k\frac{d\hat{T}}{dx}\frac{dN_i}{dx}dx = \int_e QN_i\,dx - q_i \qquad \text{for } i = 1,2 \qquad (2.2.6a)$$

This is the one-dimensional form of equation (1.7.3), in which (a), the term which is premultiplied by Ψ, has been omitted, on the understanding that the boundary condition on Γ_1 will be exactly satisfied later and (b) the integrals over the boundaries have been reduced to the summation of point values, i.e.,

$$q_i = [\hat{q}N_i]_\Gamma = [\hat{q}N_i]_{\Gamma_1} + [\hat{q}N_i]_{\Gamma_2} \qquad (2.2.6b)$$

Here $\hat{q}$ is unknown on Γ_1 and $\hat{q} = \aleph$ on Γ_2. Inserting the assumed representation for $\hat{T}$ from equation (2.2.5) leads to the pair of equations

$$\sum_{j=1}^{2}\left[\int_e k\frac{dN_j}{dx}\frac{dN_i}{dx}dx\right]T_j = \int_e QN_i\,dx - q_i \qquad i = 1,2 \qquad (2.2.7)$$

and, with the function Q linearly interpolated in terms of its nodal values, as in equation (1.7.9), these equations can be written as

$$\sum_{j=1}^{2}\left[\int_e k\frac{dN_j}{dx}\frac{dN_i}{dx}dx\right]T_j = \sum_{j=1}^{2}\left[\int_e N_jN_i\,dx\right]Q_j - q_i \qquad i = 1,2 \quad (2.2.8)$$

Since the shape functions are defined in terms of the coordinate ξ, it is convenient to apply the mapping of (2.2.4a) and re-write these equations as

$$\sum_{j=1}^{2}\left[\int_{-1}^{1} k\frac{dN_j}{d\xi}\frac{d\xi}{dx}\frac{dN_i}{d\xi}\frac{d\xi}{dx}\frac{dx}{d\xi}d\xi\right]T_j = \sum_{j=1}^{2}\left[\int_{-1}^{1} N_j N_i\frac{dx}{d\xi}d\xi\right]Q_j - q_i \qquad i = 1, 2$$

(2.2.9)

The mapping implies that the relationship

$$dx = h_e\, d\xi/2$$

(2.2.10)

holds for this example, so that the finite element approximation is determined from the solution of the equations

$$\sum_{j=1}^{2}\left[\frac{2}{h_e}\int_{-1}^{1} k\frac{dN_j}{d\xi}\frac{dN_i}{d\xi}d\xi\right]T_j = \sum_{j=1}^{2}\left[\frac{h_e}{2}\int_{-1}^{1} N_j N_i\, d\xi\right]Q_j - q_i \qquad i = 1, 2$$

(2.2.11)

This pair of equations may be expressed in the matrix form

$$\mathbf{K}_e\mathbf{T}_e = \mathbf{M}_e\mathbf{Q}_e - \mathbf{q}_e$$

(2.2.12)

The 2×2 matrices $\mathbf{K}_e$ and $\mathbf{M}_e$ are the one-dimensional linear element stiffness and mass matrices respectively, with typical entries

$$[\mathbf{K}_e]_{ij} = \frac{2}{h_e}\int_{-1}^{1} k\frac{dN_j}{d\xi}\frac{dN_i}{d\xi}d\xi$$

(2.2.13)

and

$$[\mathbf{M}_e]_{ij} = \frac{h_e}{2}\int_{-1}^{1} N_j N_i\, d\xi$$

(2.2.14)

The element vectors $\mathbf{T}_e, \mathbf{Q}_e$ and $\mathbf{q}_e$ are given by

$$\mathbf{T}_e = \begin{bmatrix} T_1 \\ T_2 \end{bmatrix} \qquad \mathbf{Q}_e = \begin{bmatrix} Q_1 \\ Q_2 \end{bmatrix} \qquad \mathbf{q}_e = \begin{bmatrix} q_1 \\ q_2 \end{bmatrix}$$

(2.2.15)

Using the definition of the shape functions given in equation (2.2.4b), the evaluation of the integrals in equations (2.2.13) and (2.2.14) is readily accomplished analytically and the element stiffness and mass matrices can be shown to be

$$\mathbf{K}_e = \frac{k_e}{h_e}\begin{bmatrix} 1 & -1 \\ -1 & 1 \end{bmatrix} \qquad \mathbf{M}_e = \frac{h_e}{6}\begin{bmatrix} 2 & 1 \\ 1 & 2 \end{bmatrix}$$

(2.2.16)

Here it has been assumed that the thermal conductivity, k, is constant and equal to k_e on the element.

In the discussion of this section, the boundary conditions have been treated in a fairly general manner. The following example illustrates how the implementation of specific boundary conditions can be incorporated within the solution procedure.

Example 2.1

Consider the example of determining the temperature distribution in the thin cylindrical rod of Figure 2.2.1, when the length of the rod $L = 1$ and the rod is constructed of a material of constant thermal conductivity $k = 10$. The end $x = 0$ of the rod is maintained at a temperature $T = 5$, while a heat flux $q = 3T - 7$ is applied at the end $x = 1$. In addition, it will be assumed that, at position x in the rod, heat is being generated at a rate $Q = 180x$ per unit length. The distribution of temperature, T, within the rod satisfies the equation

$$10\frac{d^2 T}{dx^2} + 180x = 0 \qquad 0 < x < 1 \tag{2.2.17}$$

with the boundary conditions

$$T = 5 \quad \text{at } x = 0; \qquad q = 3T - 7 \quad \text{at } x = 1 \tag{2.2.18}$$

In terms of the general boundary conditions of equations (2.2.2) and (2.2.3), we see that in this case Γ_1 is the point $x = 0$, Γ_2 is the point $x = 1$, $f = 5$ and $\aleph = 3T - 7$. Using a single linear element to represent the region $0 \leqslant x \leqslant 1$, with node 1 at $x = 0$ and node 2 at $x = 1$, we have that $h_e = 1$ and the element equations (2.2.12) becomes

$$10\begin{bmatrix} 1 & -1 \\ -1 & 1 \end{bmatrix}\begin{bmatrix} T_1 \\ T_2 \end{bmatrix} = \frac{1}{6}\begin{bmatrix} 2 & 1 \\ 1 & 2 \end{bmatrix}\begin{bmatrix} Q_1 \\ Q_2 \end{bmatrix} - \begin{bmatrix} q_1 \\ q_2 \end{bmatrix} \tag{2.2.19a}$$

where, by definition, $Q_1 = 0$ and $Q_2 = 180$. Since $[N_1]_{\Gamma_1} = 1$, $[N_1]_{\Gamma_2} = 0$, $[N_2]_{\Gamma_1} = 0$, $[N_2]_{\Gamma_2} = 1$, then

$$q_1 = [\dot{q}]_{x=0} \qquad q_2 = [\aleph]_{x=1} = \aleph_2 \tag{2.2.19b}$$

and the equation corresponding to node 2 can be written as

$$10(-T_1 + T_2) = 360/6 - \aleph_2 \tag{2.2.20}$$

Inserting the boundary conditions $T_1 = 5$, $\aleph_2 = 3T_2 - 7$ into this equation, gives the result $T_2 = 9$. The equation corresponding to node 1 is

$$10(T_1 - T_2) = 180/6 - q_1 \tag{2.2.21}$$

and this can be employed, since now both T_1 and T_2 are known, to deduce that $q_1 = 70$.

This simple problem is amenable to analytical solution and the exact distribution of temperature is found to be $T = 5 + 7x - 3x^3$. It is easily checked that, in this case, the Galerkin finite element method has reproduced the exact nodal values for the unknowns T_2 and q_1, although the solution between the nodes is obviously not exact (see Figure 2.2.3).

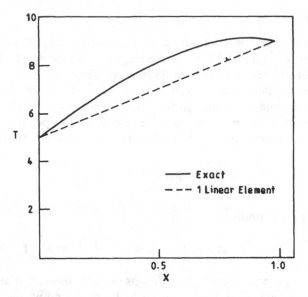

Figure 2.2.3 *Solution using one linear element.*

2.2.2   The quadratic element

If, using a single element, we wish to produce a more accurate solution to our problem, over the whole range of values of x, then we must employ shape functions of a higher polynomial order. This process is illustrated by considering how the problem formulated in equations (2.2.1)–(2.2.3) may be solved by using a single quadratic element. Nodes, numbered 1, 2 and 3, are located on the quadratic element e as shown in Figure 2.2.4. Using the quadratic element shape functions, defined from equation (1.5.14) as

$$N_1(\xi) = -\xi(1-\xi)/2 \qquad N_2(\xi) = (1-\xi^2) \qquad N_3(\xi) = \xi(1+\xi)/2 \qquad (2.2.22)$$

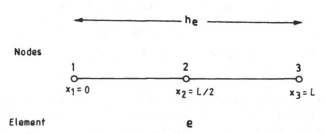

Figure 2.2.4 *Node numbers for a quadratic element.*

for $-1 \leqslant \xi \leqslant 1$, the standard three-noded quadratic element of Section 1.5.2, can be mapped into the element e via the isoparametric mapping

$$x = \sum_{j=1}^{3} x_j N_j(\xi) = x_2 + h_e \xi/2 + (x_1 - 2x_2 + x_3)\xi^2/2 = h_e/2 + h_e \xi/2 \quad (2.2.23)$$

It can be observed that the quadratic term in the mapping vanishes when, as in this case, the interior node 2 is centrally located on the element e. Again following the procedure outlined in Section 1.7, we now work with trial functions

$$\hat{T} = \sum_{j=1}^{3} T_j N_j(\xi) \quad (2.2.24)$$

and the finite element approximation is determined from the requirement that

$$\int_e k \frac{d\hat{T}}{dx} \frac{dN_i}{dx} dx = \int_e Q N_i \, dx - q_i \quad \text{for } i = 1, 2, 3 \quad (2.2.25)$$

Inserting the assumed form of $\hat{T}$ from equation (2.2.24), leads to the set of three equations

$$\sum_{j=1}^{3} \left[\int_e k \frac{dN_j}{dx} \frac{dN_i}{dx} dx \right] T_j = \int_e Q N_i \, dx - q_i \quad i = 1, 2, 3 \quad (2.2.26)$$

When Q is quadratically interpolated in terms of its nodal values, the mapping of equation (2.2.23) can be used to evaluate both integrals. The result is that these equations may again be expressed in the matrix form

$$\mathbf{K}_e \mathbf{T}_e = \mathbf{M}_e \mathbf{Q}_e - \mathbf{q}_e \quad (2.2.27)$$

where the 3×3 matrices $\mathbf{K}_e$ and $\mathbf{M}_e$ are now the one-dimensional quadratic element stiffness and mass matrices respectively. The typical entries in these matrices are defined by exactly the same expressions which were produced for the corresponding linear element matrices in equations (2.2.13) and (2.2.14), viz

$$[\mathbf{K}_e]_{ij} = \frac{2}{h_e} \int_{-1}^{1} k \frac{dN_j}{d\xi} \frac{dN_i}{d\xi} d\xi \qquad [\mathbf{M}_e]_{ij} = \frac{h_e}{2} \int_{-1}^{1} N_j N_i \, d\xi \quad (2.2.28)$$

For this element, the vectors $\mathbf{T}_e, \mathbf{Q}_e$ and $\mathbf{q}_e$ are defined to be

$$\mathbf{T}_e = \begin{bmatrix} T_1 \\ T_2 \\ T_3 \end{bmatrix} \qquad \mathbf{Q}_e = \begin{bmatrix} Q_1 \\ Q_2 \\ Q_3 \end{bmatrix} \qquad \mathbf{q}_e = \begin{bmatrix} q_1 \\ 0 \\ q_3 \end{bmatrix} \quad (2.2.29)$$

with $q_2 = 0$ since $[N_2]_{\Gamma_4} = [N_2]_{\Gamma_2} = 0$. Using the expressions for the shape functions given in equation (2.2.22), the entries in the element mass and stiffness matrices can be determined, assuming that the thermal conductivity $k = k_e$ of the element is constant, by performing analytically the integrations in equation

(2.2.28). The result is that

$$\mathbf{K}_e = \frac{k_e}{3h_e} \begin{bmatrix} 7 & -8 & 1 \\ -8 & 16 & -8 \\ 1 & -8 & 7 \end{bmatrix} \qquad \mathbf{M}_e = \frac{h_e}{30} \begin{bmatrix} 4 & 2 & -1 \\ 2 & 16 & 2 \\ -1 & 2 & 4 \end{bmatrix} \qquad (2.2.30)$$

The imposition of typical boundary conditions of interest within this approach will again be illustrated by means of an example.

Example 2.2

We reconsider the problem of Example 2.1 and produce a solution using a single quadratic element. With the representation of the region as shown in Figure 2.2.4, we have that $L = h_e = 1$, $q_1 = [\hat{q}]_{x=0}$, $q_3 = \aleph_3$ and the equations corresponding to nodes 2 and 3 are,

$$10[-8T_1 + 16T_2 - 8T_3]/3 = \tfrac{180}{30}[2(0) + 16(0.5) + 2(1)]$$
$$10[T_1 - 8T_2 + 7T_3]/3 = \tfrac{180}{30}[-1(0) + 2(0.5) + 4(1)] - \aleph_3 \qquad (2.2.31)$$

Inserting the boundary conditions $T_1 = 5$ and $\aleph_3 = 3T_3 - 7$, leads to the solution $T_2 = 8.125$ and $T_3 = 9$. When these values have been obtained, the equation corresponding to node 1 can be used to determine q_1 as

$$q_1 = -\tfrac{10}{3}[7T_1 - 8T_2 + T_3] + \tfrac{180}{30}[2/2 - 1] = 70 \qquad (2.2.32)$$

The computed nodal values are again exact. A comparison between the exact

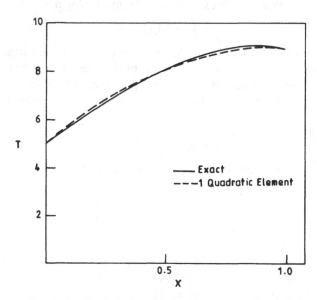

Figure 2.2.5   *Solution using one quadratic element.*

and the quadratic finite element solution over the whole element is shown in Figure 2.2.5. The improvement in the solution accuracy over that attained by the use of a single linear element, shown in Figure 2.2.3, is clearly apparent.

2.2.3 The use of numerical integration

For the examples which have been considered, exact evaluation of all integrals has been readily performed. However, an alternative approach to the evaluation of these integrals, which is well-suited for computer implementation, is the use of numerical integration techniques. To illustrate the use of such methods, we first need to recognise that, following the application of an isoparametric mapping, the integrals of interest here can generally be expressed in the form

$$\int_e g(x)\,dx = \int_{-1}^{1} G(\xi)\,d\xi \tag{2.2.33}$$

Integrals over the range $-1 \leqslant \xi \leqslant 1$ are readily approximated using Gauss–Legendre numerical integration formulae. These formulae require the evaluation, or sampling, of the integrand at n_s sampling points $\xi_1, \xi_2, \ldots, \xi_{n_s}$ and can be expressed as

$$\int_e g(x)\,dx = \int_{-1}^{1} G(\xi)\,d\xi \approx \sum_{j=1}^{n_s} \omega_j G(\xi_j) \tag{2.2.34}$$

where ω_j denotes the weight applicable to the sampling point ξ_j. With correct choice of the sampling points and of the weights, this formula integrates exactly any polynomial of degree $2n_s - 1$. The values

$$\begin{aligned} \xi_1 &= -1/\sqrt{3} & \xi_2 &= 1/\sqrt{3} \\ \omega_1 &= 1 & \omega_2 &= 1 \end{aligned} \tag{2.2.35}$$

are employed when two sampling points are used (i.e. $n_s = 2$) and the corresponding formula is exact when G is any cubic function of ξ. The values

$$\begin{aligned} \xi_1 &= -\sqrt{3/5} & \xi_2 &= 0 & \xi_2 &= \sqrt{3/5} \\ \omega_1 &= 5/9 & \omega_2 &= 8/9 & \omega_3 &= 5/9 \end{aligned} \tag{2.2.36}$$

are appropriate when the choice $n_s = 3$ is adopted and this leads to a formula which is exact when G is any quintic function of ξ. The location of the sampling points, and the correct values for the corresponding weights, for other choices of n_s may be found in standard texts on numerical analysis.

Note that, for the examples considered above, the exact evaluation of the entries in the stiffness and mass matrices for the linear element can be achieved with the choice $n_s = 2$, while the choice $n_s = 3$ must be made to ensure exact evaluation for the quadratic element.

2.3   ONE-DIMENSIONAL PROBLEM SOLVED USING AN ASSEMBLY OF ELEMENTS

In principle, the accuracy of the approximation to the solution of the one-dimensional heat conduction problem, described in the previous section, can be improved by using elements of successively higher order. In practice, however, the approach which is generally adopted by the analyst is to attempt to attain greater accuracy by subdividing the domain into a number of low-order elements. The approximate solution will then be a continuous function which is defined in terms of a local (or piecewise) low-order polynomial over each element. With this approach, the approximate equations can be formed for each element in turn by employing exactly the same procedures as have been followed previously for the single element. Then, provided that these equations are combined in the correct fashion, the resulting equations may be solved to produce the required solution over the whole domain.

2.3.1   The use of linear elements

Consider, for example, the use of M non-overlapping two-noded linear elements to represent the domain $0 \leqslant x \leqslant L$. For convenience, it will be assumed here that the nodes and the elements are numbered as shown in Figure 2.3.1. A typical element e has nodes that can be locally numbered 1 and 2 and which are globally numbered I and J, say, respectively. Our basic linear element, defined on the

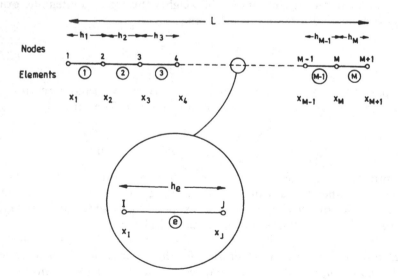

Figure 2.3.1   *Subdivision of a domain using M linear elements.*

region $-1 \leqslant \xi \leqslant 1$, can be mapped into this element e by the mapping function

$$x = \sum_{j=1}^{2} x_{je} N_j(\xi) = x_I + h_e/2 + h_e \xi/2 \qquad (2.3.1)$$

where $N_1(\xi)$ and $N_2(\xi)$ are the standard linear finite element shape functions defined in equation (2.2.4b), $x_{1e} = x_I, x_{2e} = x_J$ and $h_e = x_J - x_I$ is the length of the element. Note now that the use of the subscript e indicates a reference to local numbering and values. On element e, we work with trial functions

$$\hat{T} = \sum_{j=1}^{2} T_{je} N_j(\xi) \qquad (2.3.2)$$

where the definition $T_{1e} = T_I$, $T_{2e} = T_J$ will ensure the continuity of the approximation as we pass from one element to the next. From equation (2.2.11), we deduce that the finite element approximation over element e is determined from the requirement that

$$\sum_{j=1}^{2} \left[\frac{2}{h_e} \int_{-1}^{1} k \frac{dN_j}{d\xi} \frac{dN_i}{d\xi} d\xi \right] T_{je} = \sum_{j=1}^{2} \left[\frac{h_e}{2} \int_{-1}^{1} N_j N_i \, d\xi \right] Q_{je} - q_{ie} \qquad i = 1, 2 \qquad (2.3.3)$$

where $Q_{1e} = Q_I, Q_{2e} = Q_J$ and $q_{1e} = q_{I(e)}, q_{2e} = q_{J(e)}$. The quantities $q_{I(e)}, q_{J(e)}$ denote the fluxes at nodes I and J respectively as those nodes are approached from within element e. This pair of equations may be expressed in the matrix form

$$\mathbf{K}_e \mathbf{T}_e = \mathbf{M}_e \mathbf{Q}_e - \mathbf{q}_e \qquad (2.3.4)$$

with the element stiffness and mass matrices $\mathbf{K}_e$ and $\mathbf{M}_e$ being exactly as given in equation (2.2.16) and the element vectors are now

$$\mathbf{T}_e = \begin{bmatrix} T_{1e} \\ T_{2e} \end{bmatrix} \qquad \mathbf{Q}_e = \begin{bmatrix} Q_{1e} \\ Q_{2e} \end{bmatrix} \qquad \mathbf{q}_e = \begin{bmatrix} q_{1e} \\ q_{2e} \end{bmatrix} \qquad (2.3.5)$$

The equations for a general element e have now been determined and the equations arising from each element in turn can be obtained from equations (2.3.3)–(2.3.5) by relating the local node numbers 1 and 2 to the actual global node numbers I and J, and using the appropriate element length, h_e, and conductivity, k_e, in each case. With the nodes numbered consecutively from left to right, as shown in Figure 2.3.1, the resulting equations take the form

Element 1

$$k_1(T_1 - T_2)/h_1 = h_1(2Q_1 + Q_2)/6 - q_{1(1)}$$
$$k_1(-T_1 + T_2)/h_1 = h_1(Q_1 + 2Q_2)/6 - q_{2(1)}$$

Element 2

$$k_2(T_2 - T_3)/h_2 = h_2(2Q_2 + Q_3)/6 - q_{2(2)}$$
$$k_2(-T_2 + T_3)/h_2 = h_2(Q_2 + 2Q_3)/6 - q_{3(3)} \qquad (2.3.6)$$

$$\vdots$$

Element M

$$k_M(T_M - T_{M+1})/h_M = h_M(2Q_M + Q_{M+1})/6 - q_{M(M)}$$

$$\frac{k_M}{h_M}(-T_M + T_{M+1}) = \frac{h_M}{6}(Q_M + 2Q_{M+1}) - q_{M+1(M)}$$

This system appears to have more unknowns than equations, but if the equations arising from each node are added and use is made of the requirement of continuity of flux, so that here

$$q_{I(e-1)} + q_{I(e)} = 0 \qquad (2.3.7)$$

for each interior node I, then we get the system

$$k_1(T_1 - T_2)/h_1 = h_1(2Q_1 + Q_2)/6 - q_1$$
$$k_1(-T_1 + T_2)/h_1 + k_2(T_2 - T_3)/h_2 = h_1(Q_1 + 2Q_2)/6 + h_2(2Q_2 + Q_3)/6$$
$$k_2(-T_2 + T_3)/h_2 + k_3(T_3 - T_4)/h_3 = h_2(Q_2 + 2Q_3)/6 + h_3(2Q_3 + Q_4)/6 \quad (2.3.8)$$
$$\vdots$$
$$k_M(T_M - T_{M+1})/h_M = h_M(Q_M + Q_{M+1})/6 - q_{M+1}$$

Following this process of assembly, it can be observed that, with the boundary conditions of equations (2.2.2) and (2.2.3), we now have a set of $M + 3$ equations which can be solved for the $M + 3$ unknowns $T_1, T_2, \ldots, T_M, T_{M+1}, q_1(= q_{1(1)})$, $q_{M+1}(= q_{M+1(M)})$. If any node is a source or a sink of heat, then the sum of the fluxes at that node will not vanish. In this case, equation (2.3.7) will be replaced by the condition

$$q_{I(e-1)} + q_{I(e)} = \delta_I \qquad (2.3.9)$$

The quantity δ_I, which is the prescribed flux for the source or the sink at node I, will appear in the assembled equation for node I.

Note also that if the region of interest is constructed of two materials, with different thermal properties, then, as we have seen from Section 1.3.2, the above approach remains applicable provided that the material interface coincides with the position of one of the nodes. The heat flux remains continuous across such an interface and all we need to ensure is that the appropriate thermal conductivity is used when constructing the individual element matrices on either side of the interface.

Example 2.3

We return to the problem of Example 2.1 and consider the construction of an approximate solution using four linear elements of equal length. We employ the node and element numbering system shown in Figure 2.3.1, with $M = 4$ and $h_1 = h_2 = h_3 = h_4 = 1/4$. The local and global node numbers for each element are then related according to

element	local node 1	local node 2	
1	1	2	
2	2	3	(2.3.10)
3	3	4	
4	4	5	

For this problem, the equation system for element e is that defined by equation (2.3.4) with the entries in the matrices and vectors which appear in this equation being defined in equations (2.3.5) and (2.2.16). Using this equation for each element in turn, and performing the assembly process, leads to the equation system

$$10(T_1 - T_2)/(1/4) = 180(1/4)[(1/4)]/6 - q_1$$
$$10(-T_1 + 2T_2 - T_3)/(1/4) = 180(1/4)[4(1/4) + 1/2]/6$$
$$10(-T_2 + 2T_3 - T_4)/(1/4) = 180(1/4)[1/4 + 4(1/2) + 3/4]/6 \quad (2.3.11)$$
$$10(-T_3 + 2T_4 - T_5)/(1/4) = 180(1/4)[1/2 + 4(3/4) + 1]/6$$
$$10(-T_4 + T_5)/(1/4) = 180(1/4)[3/4 + 2]/6 - q_5$$

The boundary conditions for this problem specify the values $T_1 = 5$ and $q_5 = \aleph_5 = 3T_5 - 7$. The unknowns are then T_2, T_3, T_4, T_5 and q_1. Inserting the prescribed boundary conditions, the equations corresponding to nodes 2–5 can be expressed in the matrix form

$$\mathbf{KT} = \mathbf{r} \quad (2.3.12a)$$

where

$$\mathbf{K} = \begin{bmatrix} 80 & -40 & 0 & 0 \\ -40 & 80 & -40 & 0 \\ 0 & -40 & 80 & -40 \\ 0 & 0 & -40 & 43 \end{bmatrix} \quad \mathbf{T} = \begin{bmatrix} T_2 \\ T_3 \\ T_4 \\ T_5 \end{bmatrix} \quad \mathbf{r} = \begin{bmatrix} 7.5 \times 1.5 + 200 \\ 7.5 \times 3 \\ 7.5 \times 4.5 \\ 7.5 \times 2.75 + 7 \end{bmatrix}$$

$$(2.3.12b)$$

This system may be solved to obtain the values $T_2 = 6.703\,125$, $T_3 = 8.125$, $T_4 = 8.984\,375$ and $T_5 = 9$. When these values have been computed, the equation corresponding to node 1 may be used to determine that $q_1 = 70$. It is readily checked that these nodal values are again exact. The solution is compared with the exact solution in Figure 2.3.2 and the improvement in the solution accuracy, over that achieved with a single linear element in Figure 2.2.3, is clearly apparent.

It is interesting to note that the equation corresponding to a typical interior node (e.g. mode 2) is

$$10(-T_1 + 2T_2 - T_3)/(1/4) = 180(1/4)[4(1/4) + 1/2]/6 \quad (2.3.13)$$

while a direct application of the finite difference method to the governing equation

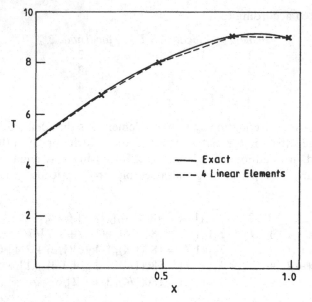

Figure 2.3.2   *Solution using 4 linear elements.*

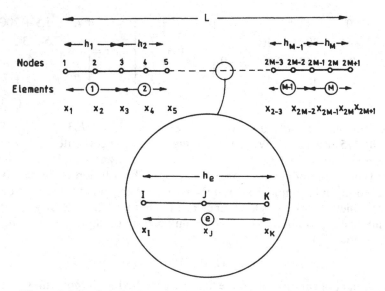

Figure 2.3.3   *Subdivision of a domain using M quadratic elements.*

at the same node would produce the discrete equation

$$10(-T_1 + 2T_2 - T_3)/(1/4)^2 = 180(1/4) \qquad (2.3.14)$$

It can be seen that this equation is identical to that produced by the finite element method in this case, as the heat generation rate Q is a linear function of x. At the boundary node 5, however, use of a standard central difference approximation produces an equation which is different from that derived from the finite element method. In fact, the finite element treatment of the flux boundary condition is more accurate than that of the finite difference method and this is demonstrated by the fact that the nodal values produced for this problem by the finite difference method are not exact. Similarly, the finite element method is more accurate than the one-sided second-order finite difference approximation in predicting the heat flux that is required at node 1 to maintain the prescribed fixed temperature at that node.

2.3.2 The use of quadratic elements

An assembly of quadratic elements can also be used to solve the problem of equations (2.2.1)–(2.2.3) in a similar fashion. Suppose that M non-overlapping three-noded quadratic elements are used to represent the domain $0 \leqslant x \leqslant L$, as shown in Figure 2.3.3. A typical element e has nodes labelled I, J and K which are locally numbered 1, 2 and 3 respectively. It will be assumed that each quadratic element is such that the interior node is centrally located on the element. The standard three-noded quadratic element of Section 1.5.2, defined over the range $-1 \leqslant \xi \leqslant 1$, can be mapped into element e by means of the isoparametric mapping

$$x = \sum_{j=1}^{3} x_{je} N_j(\xi) = x_I + h_e/2 + h_e \xi/2 \qquad (2.3.15)$$

since $x_{1e} = x_I, x_{2e} = x_J = x_I + h_e/2$, $x_{3e} = x_K = x_I + h_e$ and where we have used the definition of the shape functions given in equation (2.2.22). Over this element, we work with trial functions

$$\hat{T} = \sum_{j=1}^{3} T_{je} N_j(\xi) \qquad (2.3.16)$$

and continuity of the approximation from element to element is assured by the choice $T_{1e} = T_I$, $T_{2e} = T_J$ and $T_{3e} = T_K$. From equations (2.2.27), (2.2.29) and (2.2.30), it is apparent that the resulting Galerkin approximation over this element will lead to the equation

$$\mathbf{K}_e \mathbf{T}_e = \mathbf{M}_e \mathbf{Q}_e - \mathbf{q}_e \qquad (2.3.17)$$

where the matrices $\mathbf{K}_e$ and $\mathbf{M}_e$ are exactly as defined in equation (2.2.30) and

$$\mathbf{T}_e = \begin{bmatrix} T_{1e} \\ T_{2e} \\ T_{3e} \end{bmatrix} \qquad \mathbf{Q}_e = \begin{bmatrix} Q_{1e} \\ Q_{2e} \\ Q_{3e} \end{bmatrix} \qquad \mathbf{q}_e = \begin{bmatrix} q_{1e} \\ 0 \\ q_{3e} \end{bmatrix} \qquad (2.3.18)$$

With the matrix form of the equation for a general element determined in this fashion, the system of elements is assembled using the global node numbering. Here it should be observed that no assembly is necessary for the interior node of each element. With the nodes numbered sequentially, as shown in Figure 2.3.3, the form of the assembled equation system is

$$k_1(7T_1 - 8T_2 + T_3)/(3h_1) = h_1(4Q_1 + 2Q_2 - Q_3)/30 - q_1$$

$$k_1(-8T_1 + 16T_2 - 8T_3)/(3h_1) = h_1(2Q_1 + 16Q_2 + 2Q_3)/30$$

$$k_1(T_1 - 8T_2 + 7T_3)/(3h_1) + k_2(7T_3 - 8T_4 + T_5)/(3h_2) = h_1(-Q_1 + 2Q_2 + 4Q_3)/30$$
$$+ h_2(4Q_3 + 2Q_4 - Q_5)/30$$

$$k_2(-8T_3 + 16T_4 - 8T_5)/(3h_2) = h_2(2Q_3 + 16Q_4 + 2Q_5)/30$$

$$k_2(T_3 - 8T_4 + 7T_5)/(3h_2) + k_3(7T_5 - 8T_6 + T_7)/(3h_3) = h_2(-Q_3 + 2Q_4 + 4Q_5)/30$$
$$+ h_3(4Q_5 + 2Q_6 - Q_7)/30$$
$$\vdots$$

$$k_M(T_{2M-1} - 8T_{2M} + 7T_{2M+1})/(3h_M) = h_M(-Q_{2M-1} + 2Q_{2M} + 4Q_{2M+1})/30 - q_{2M+1}$$

$$(2.3.19)$$

The process of boundary condition application is again best illustrated in terms of a specific example.

Example 2.4

We reconsider the problem defined in Example 2.1 and produce a solution using two quadratic elements of equal length. The mesh used is shown in Figure 2.3.4. The interior nodes are centrally located on the elements. It can be observed that this mesh has the same number, and the same distribution, of nodes as the mesh of four linear elements which was used in Example 2.3. However, in this case, the local and global node numbers for each element are now related

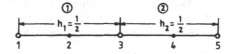

Figure 2.3.4 *Quadratic elements of equal length.*

according to

element	local node 1	local node 2	local node 3	
1	1	2	3	(2.3.20)
2	3	4	5	

The matrix form of the equation for a typical element is again equation (2.2.27), with the element matrices K_e and M_e now determined as

$$K_e = \frac{10}{3h_e}\begin{bmatrix} 7 & -8 & 1 \\ -8 & 16 & -8 \\ 1 & -8 & 7 \end{bmatrix} \qquad M_e = \frac{h_e}{30}\begin{bmatrix} 4 & 2 & -1 \\ 2 & 16 & 2 \\ -1 & 2 & 4 \end{bmatrix} \qquad (2.3.21a)$$

The element vectors T_e and q_e are as defined in equation (2.3.18) and, for this problem,

$$Q_e = 180\begin{bmatrix} x_{1e} \\ x_{2e} \\ x_{3e} \end{bmatrix} \qquad (2.3.21b)$$

Using the element equation for elements 1 and 2 in turn, and performing the element assembly gives, since $h_1 = h_2 = 1/2$, the equation system

$$10[7T_1 - 8T_2 + T_3]/3(1/2) = 180(1/2)[2(1/4) - 1/2]/30 - q_1$$
$$10[-8T_1 + 16T_2 + 8T_3]/3(1/2) = 180(1/2)[16(1/4) + 2(1/2)]/30$$
$$10[T_1 - 8T_2 + 7T_3]/3(1/2) + 10[7T_3 - 8T_4 + T_5]/3(1/2) \qquad (2.3.22)$$
$$= 180(1/2)[2(1/4) + 4(1/2)]/30 + 180(1/2)[2(1/4) + 2(3/4) + 1]/30$$
$$10[-8T_3 + 16T_4 - 8T_5]/3(1/2) = 180(1/2)[2(1/2) + 16(3/4) + 2]/30$$
$$10[T_3 - 8T_4 + 7T_5]/3(1/2) = 180(1/2)[-1/2 + 2(3/4) + 4]/30 - q_5$$

The boundary conditions for this problem require that $T_1 = 5$ and $q_5 = 3T_5 - 7$. When these values are inserted into this equation set, the equations corresponding to nodes 2–5 can be written in the matrix form

$$KT = r \qquad (2.3.23)$$

where

$$K = \frac{20}{3}\begin{bmatrix} 16 & -8 & 0 & 0 \\ -8 & 14 & -8 & 1 \\ 0 & -8 & 16 & -8 \\ 0 & 1 & -8 & 7+3(3/20) \end{bmatrix} \qquad (2.3.24a)$$

and

$$T = \begin{bmatrix} T_2 \\ T_3 \\ T_4 \\ T_5 \end{bmatrix}, \qquad r = \begin{bmatrix} 15 + 800/3 \\ 15 - 100/3 \\ 45 \\ 15 + 7 \end{bmatrix} \qquad (2.3.24b)$$

The solution of this equation system produces the values $T_2 = 6.703\,125$, $T_3 = 8.125$, $T_4 = 8.984\,373$ and $T_5 = 9$. The result $q_1 = 70$ follows from using the equation corresponding to node 1. This solution is again nodally exact and is an improvement over that obtained using four linear elements in that it is graphically almost indistinguishable from the exact solution over the whole region.

2.3.3   The use of numerical integration

When the region of interest is represented by an assembly of elements, there are no additional complications introduced if numerical integration procedures are employed in the evaluation of the element integrals. The integrations are performed for each element in turn, using the appropriate integration formula, as shown in equations (2.2.34)–(2.2.36). The numerically evaluated contributions are then assembled in exactly the same manner as that described above.

2.4   A TWO-DIMENSIONAL PROBLEM SOLVED USING A SINGLE ELEMENT

Consider the problem of steady heat conduction in a two-dimensional region, Ω, with closed boundary curve, Γ, and of thermal conductivity k. If heat is generated within the region at a prescribed rate Q per unit area, then the distribution of temperature through Ω is governed by the solution of the equation

$$\text{div}(k\,\text{grad}\,T) + Q = \frac{\partial}{\partial x}\left[k\frac{\partial T}{\partial x}\right] + \frac{\partial}{\partial y}\left[k\frac{\partial T}{\partial y}\right] + Q = 0 \qquad \text{in } \Omega \qquad (2.4.1)$$

subject to the general boundary conditions

$$T = f(\mathbf{x}) \qquad \text{for } \mathbf{x} \text{ on } \Gamma_1 \tag{2.4.2}$$

$$q = -(k\,\text{grad}\,T)\cdot\mathbf{n} = -k\frac{\partial T}{\partial n} = \aleph(\mathbf{x}, T) \qquad \text{for } \mathbf{x} \text{ on } \Gamma_2 \tag{2.4.3}$$

We recall that $\Gamma_1 \cup \Gamma_2 = \Gamma$, $\Gamma_1 \cap \Gamma_2 = 0$ and that $f, \aleph$ will be prescribed functions. As in the one-dimensional case considered previously, the application of particular boundary conditions can be achieved by appropriate definition of Γ_1 and Γ_2 and the specification of the values of f and $\aleph$.

2.4.1   The four-noded rectangular element

Suppose, initially, that the domain Ω is in the form of a rectangular defined by $0 \leqslant x \leqslant L_x, 0 \leqslant y \leqslant L_y$, as shown in Figure 2.4.1(a). If nodes numbered 1, 2, 3 and

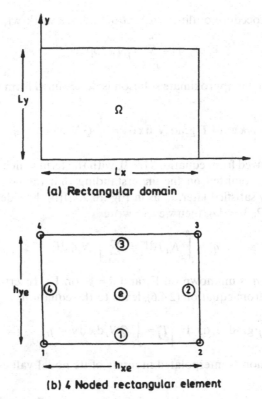

(a) Rectangular domain

(b) 4 Noded rectangular element

Figure 2.4.1 *Rectangular domain and four-noded rectangular element.*

4 are located at the corners of the region, the standard four-noded square element of Section 1.6.1, defined on the range $-1 \leqslant \xi, \eta \leqslant 1$, can be mapped into an element e which exactly represents the domain Ω, as shown in Figure 2.4.1(b). The element shape functions are given from equation (1.6.8) as

$$N_1 = (1 - \xi)(1 - \eta)/4 \qquad N_2 = (1 + \xi)(1 - \eta)/4$$
$$N_3 = (1 + \xi)(1 + \eta)/4 \qquad N_4 = (1 - \xi)(1 + \eta)/4 \qquad -1 \leqslant \xi, \eta \leqslant 1 \quad (2.4.4)$$

with the isoparametric mapping of equation (1.6.12) used to relate the (x, y) and the (ξ, η) coordinate systems according to

$$x = \sum_{j=1}^{4} x_j N_j(\xi, \eta) = h_{xe}/2 + h_{xe}\xi/2$$

$$y = \sum_{j=1}^{4} y_j N_j(\xi, \eta) = h_{ye}/2 + h_{ye}\eta/2 \qquad (2.4.5)$$

since, in this case, $x_1 = x_4 = y_1 = y_2 = 0$, $x_2 = x_3 = h_{xe} = L_x$ and $y_3 = y_4 = h_{ye} = L_y$.

Following the procedure outlined in Section 1.7, we work with trial functions

$$\hat{T} = \sum_{j=1}^{4} T_j N_j(\xi, \eta) \tag{2.4.6}$$

and the finite element approximate solution is determined from the requirement that

$$\int_e k \, \text{grad} \, \hat{T} \cdot \text{grad} \, N_i \, dx \, dy = \int_e Q N_i \, dx \, dy - q_i \tag{2.4.7}$$

This has been derived from equation (1.7.3), with the term which is premultiplied by Ψ having been omitted on the understanding that the condition $\hat{T} = \hat{f}$ on Γ_1 will be exactly satisfied later. If, as in Figure 2.4.1(b) the sides of the element are numbered 1, 2, 3 and 4, then we can write

$$q_i = \int_\Gamma N_i \hat{q} \, d\Gamma = \sum_{s=1}^{4} \int_s N_i \hat{q} \, d\Gamma \tag{2.4.8}$$

where, as before, $\hat{q}$ is unknown on Γ_1 and $\hat{q} = \aleph$ on Γ_2. Inserting the assumed expression for $\hat{T}$ from equation (2.4.6), leads to the equations

$$\sum_{j=1}^{4} \left[\int_e k \, \text{grad} \, N_j \cdot \text{grad} \, N_i \, dx \, dy \right] T_j = \int_e Q N_i \, dx \, dy - q_i \qquad i = 1, 2, 3, 4 \tag{2.4.9}$$

or, with the function Q interpolated in terms of its nodal values as in equation (1.7.9),

$$\sum_{j=1}^{4} \left[\int_e k \, \text{grad} \, N_j \cdot \text{grad} \, N_i \, dx \, dy \right] T_j = \sum_{j=1}^{4} \left[\int_e N_j N_i \, dx \, dy \right] Q_j - q_i \qquad i = 1, 2, 3, 4$$

$$\tag{2.4.10}$$

As in the one-dimensional case, we re-write these integrals, using the mapping of equation (2.4.5), in the form

$$\frac{1}{h_{xe} h_{ye}} \sum_{j=1}^{4} \left[\int_{-1}^{1} \int_{-1}^{1} k \left(h_{ye}^2 \frac{\partial N_j}{\partial \xi} \frac{\partial N_i}{\partial \xi} + h_{xe}^2 \frac{\partial N_j}{\partial \eta} \frac{\partial N_i}{\partial \eta} \right) d\xi \, d\eta \right] T_j$$

$$= \frac{h_{xe} h_{ye}}{4} \sum_{j=1}^{4} \left[\int_{-1}^{1} \int_{-1}^{1} N_j N_i \, d\xi \, d\eta \right] Q_j - q_i \qquad i = 1, 2, 3, 4 \tag{2.4.11}$$

This set of four equations may be expressed in the matrix form

$$\mathbf{K}_e \mathbf{T}_e = \mathbf{M}_e \mathbf{Q}_e - \mathbf{q}_e \tag{2.4.12}$$

The 4×4 matrices $\mathbf{K}_e$ and $\mathbf{M}_e$ are the stiffness and mass matrices respectively for the two-dimensional rectangular bilinear element, with typical entries

$$[\mathbf{K}_e]_{ij} = \frac{1}{h_{xe} h_{ye}} \left[\int_{-1}^{1} \int_{-1}^{1} k \left(h_{ye}^2 \frac{\partial N_j}{\partial \xi} \frac{\partial N_i}{\partial \xi} + h_{xe}^2 \frac{\partial N_j}{\partial \eta} \frac{\partial N_i}{\partial \eta} \right) d\xi \, d\eta \right] \tag{2.4.13}$$

and

$$[M_e]_{ij} = \frac{h_{xe}h_{ye}}{4}\left[\int_{-1}^{1}\int_{-1}^{1} N_j N_i \, d\xi \, d\eta\right] \tag{2.4.14}$$

In this case, the vectors T_e, Q_e and q_e are given by

$$T_e = \begin{bmatrix} T_1 \\ T_2 \\ T_3 \\ T_4 \end{bmatrix} \quad Q_e = \begin{bmatrix} Q_1 \\ Q_2 \\ Q_3 \\ Q_4 \end{bmatrix} \quad q_e = \begin{bmatrix} q_1 \\ q_2 \\ q_3 \\ q_4 \end{bmatrix} \tag{2.4.15}$$

Using the definition of the shape functions given in equation (2.4.4), the evaluation of the integrals in equations (2.4.13) and (2.4.14) may be accomplished analytically and the element stiffness and mass matrices determined as

$$K_e = \frac{k_e}{3h_{xe}h_{ye}} \begin{bmatrix} h_{xe}^2 + h_{ye}^2 & h_{xe}^2/2 - h_{ye}^2 & -(h_{xe}^2 + h_{ye}^2)/2 & -h_{xe}^2 + h_{ye}^2/2 \\ h_{xe}^2/2 - h_{ye}^2 & h_{xe}^2 + h_{ye}^2 & -h_{xe}^2 + h_{ye}^2/2 & -(h_{xe}^2 + h_{ye}^2)/2 \\ -(h_{xe}^2 + h_{ye}^2)/2 & -h_{xe}^2 + h_{ye}^2/2 & h_{xe}^2 + h_{ye}^2 & h_{xe}^2/2 - h_{ye}^2 \\ -h_{xe}^2 + h_{ye}^2/2 & -(h_{xe}^2 + h_{ye}^2)/2 & h_{xe}^2/2 - h_{ye}^2 & h_{xe}^2 + h_{ye}^2 \end{bmatrix} \tag{2.4.16}$$

and

$$M_e = \frac{h_{xe}h_{ye}}{36} \begin{bmatrix} 4 & 2 & 1 & 2 \\ 2 & 4 & 2 & 1 \\ 1 & 2 & 4 & 2 \\ 2 & 1 & 2 & 4 \end{bmatrix} \tag{2.4.17}$$

Here it has been assumed that the thermal conductivity of the element is constant and equal to k_e. As before, it is best to look at a particular example to illustrate how specific boundary conditions can be incorporated in practice.

Example 2.5

Consider the problem of determining the distribution of temperature in a square region $0 \leqslant x, y \leqslant 1$ of material of constant thermal conductivity $k = 2$ when heat is being generated within the region at a rate $Q = 4(x^2 + y^2)$ per unit area. The sides $x = 0$ and $y = 0$ of the region are maintained at a temperature $T = 10$, while a heat flux $q = -4T + 40$ is prescribed on the sides $x = 1$ and $y = 1$. The distribution of temperature in the region satisfies the equation

$$2\left[\frac{\partial^2 T}{\partial x^2} + \frac{\partial^2 T}{\partial y^2}\right] + 4(x^2 + y^2) = 0 \qquad 0 \leqslant x, y \leqslant 1 \tag{2.4.18}$$

and the boundary conditions

$$T = 10 \qquad \text{on } x = 0 \text{ and on } y = 0 \tag{2.4.19}$$

$$q = -4T + 40 \qquad \text{on } x = 1 \text{ and on } y = 1 \tag{2.4.20}$$

We use a single rectangular element to represent the region $0 \leqslant x, \ y \leqslant 1$, with the four nodes and the four sides numbered and located as shown in Figure 2.4.2, so that $h_{xe} = h_{ye} = 1$. In terms of the general boundary conditions of equations (2.4.2) and (2.4.3), we see that in this case Γ_1 consists of side 1 of the boundary which joins node 1 and node 2 plus side 4 which joins node 4 with node 1. Similarly, Γ_2 consists of side 2, which joins nodes 2 and 3, and side 3, which joins nodes 3 and 4. The given boundary conditions imply that $f = 10$ on Γ_1 and $\aleph = -4T + 40$ on Γ_2. The element equations (2.4.12) for this problem may then be written as

$$
\frac{2}{3}
\begin{bmatrix}
2 & -1/2 & -1 & -1/2 \\
-1/2 & 2 & -1/2 & -1 \\
-1 & -1/2 & 2 & -1/2 \\
-1/2 & -1 & -1/2 & 2
\end{bmatrix}
\begin{bmatrix}
T_1 \\ T_2 \\ T_3 \\ T_4
\end{bmatrix}
= \frac{1}{36}
\begin{bmatrix}
4 & 2 & 1 & 2 \\
2 & 4 & 2 & 1 \\
1 & 2 & 4 & 2 \\
2 & 1 & 2 & 4
\end{bmatrix}
\begin{bmatrix}
Q_1 \\ Q_2 \\ Q_3 \\ Q_4
\end{bmatrix}
-
\begin{bmatrix}
q_1 \\ q_2 \\ q_3 \\ q_4
\end{bmatrix}
\quad (2.4.21)
$$

where, by definition,

$$
q_1 = \int_4 \hat{q} N_1 \, d\Gamma + \int_1 \hat{q} N_1 \, d\Gamma \qquad q_2 = \int_1 \hat{q} N_2 \, d\Gamma + \int_2 \aleph N_2 \, d\Gamma
$$
$$
q_3 = \int_2 \aleph N_3 \, d\Gamma + \int_3 \aleph N_3 \, d\Gamma \qquad q_4 = \int_3 \aleph N_4 \, d\Gamma + \int_4 \hat{q} N_4 \, d\Gamma
\quad (2.4.22)
$$

These line integrals are best evaluated by mapping each side in turn into the standard two-noded linear element, defined on the range $-1 \leqslant \xi \leqslant 1$. To illustrate this procedure, consider a side s of an element which connects two nodes, which are globally numbered I and J, as shown in Figure 2.4.3. From equation (2.4.22)

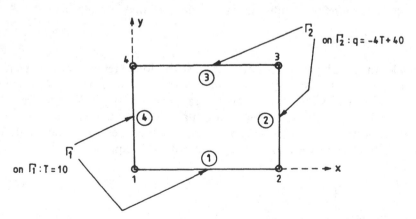

Figure 2.4.2 *Node and side numbering and boundary conditions for one four-noded element used in Example 2.5.*

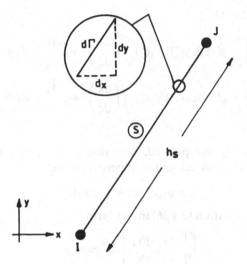

Figure 2.4.3   *Element side s of a boundary connecting two nodes I and J.*

it is apparent that the integrals which need to be evaluated over side s are of the form

$$\int_s g(x, y)N_I \, d\Gamma \quad \text{and} \quad \int_s g(x, y)N_J \, d\Gamma \tag{2.4.23}$$

Suppose that the length of the side is h_s and consider the mapping

$$x = \sum_{j=1}^{2} x_{js}N_j(\xi) = (x_I + x_J)/2 + (x_J - x_I)\xi/2$$

$$y = \sum_{j=1}^{2} y_{js}N_j(\xi) = (y_I + y_J)/2 + (y_J - y_I)\xi/2 \tag{2.4.24}$$

where $N_1(\xi)$ and $N_2(\xi)$ are the standard linear finite element shape functions defined in equation (2.2.4a), $x_{1s} = x_I$ and $x_{2s} = x_J$. This represents a mapping of a standard two-noded linear element, with nodes locally numbered 1 and 2, into the side s. Note that the use of the subscript s indicates a reference to numbering and values which are local to the side. The element length on the side may be computed as

$$d\Gamma = \sqrt{\left(\frac{dx}{d\xi}\right)^2 + \left(\frac{dy}{d\xi}\right)^2} \, d\xi \tag{2.4.25}$$

and the integrals of equation (2.4.23) can be expressed in terms of the variable

ξ, using equation (2.4.24), as

$$\int_s g(x, y)N_I \, d\Gamma = \int_{-1}^{1} g(x(\xi), y(\xi))N_1(\xi)\sqrt{\left(\frac{dx}{d\xi}\right)^2 + \left(\frac{dy}{d\xi}\right)^2} \, d\xi = \frac{h_s}{2}\int_{-1}^{1} gN_1(\xi) \, d\xi$$

$$\int_s g(x, y)N_J \, d\Gamma = \int_{-1}^{1} g(x(\xi), y(\xi))N_2(\xi)\sqrt{\left(\frac{dx}{d\xi}\right)^2 + \left(\frac{dy}{d\xi}\right)^2} \, d\xi = \frac{h_s}{2}\int_{-1}^{1} gN_2(\xi) \, d\xi$$

$$(2.4.26)$$

Further, if g is linearly interpolated, with respect to ξ, over the side in terms of its nodal values, i.e. if we adopt the approximation

$$g \approx g_{1s}N_1(\xi) + g_{2s}N_2(\xi) \tag{2.4.27}$$

then we may write equation (2.4.26) in the form

$$\int_s \begin{bmatrix} g(x, y)N_I \\ g(x, y)N_J \end{bmatrix} d\Gamma \approx \mathbf{M}_s \mathbf{g}_s \tag{2.4.28}$$

where $\mathbf{M}_s$ is the 2×2 side mass matrix whose entries are defined by

$$[\mathbf{M}_s]_{ij} = \frac{h_s}{2}\int_{-1}^{1} N_j N_i \, d\xi \tag{2.4.29a}$$

and the side vector $\mathbf{g}_s$ is given by

$$\mathbf{g}_s = \begin{bmatrix} g_{1s} \\ g_{2s} \end{bmatrix} = \begin{bmatrix} g_I \\ g_J \end{bmatrix} \tag{2.4.29b}$$

Using the results of equation (2.2.16), it follows that the required integrals can be evaluated as

$$\int_s \begin{bmatrix} g(x, y)N_I \\ g(x, y)N_J \end{bmatrix} d\Gamma = \frac{h_s}{6}\begin{bmatrix} 2g_I + g_J \\ g_I + 2g_J \end{bmatrix} \tag{2.4.30}$$

The integrals of interest for a general side s have now been determined and these can be used to evaluate the integrals in equation (2.4.22), by using the appropriate global node numbers and side length in each case. In this example, the local and global node numbers for each side are related according to

side	local node 1	local node 2
1	1	2
2	2	3
3	3	4
4	4	1

$$(2.4.31)$$

It follows, on assembling these side contributions, that

$$q_3 = \tfrac{1}{6}(\aleph_2 + 2\aleph_3) + \tfrac{1}{6}(2\aleph_3 + \aleph_4) \tag{2.4.32}$$

since the sides here are all of unit length.

The problem boundary conditions define the values

$$\aleph_2 = -4T_2 + 40 \qquad \aleph_3 = -4T_3 + 40 \qquad \aleph_4 = -4T_4 + 40 \qquad (2.4.33)$$

From the equation set (2.4.21), the equation corresponding to node 3 can then be expressed in the form

$$\begin{aligned}
\tfrac{2}{3}[-T_1 - T_2/2 + 2T_3 - T_4/2] &= \tfrac{1}{36}[Q_1 + 2Q_2 + 4Q_3 + 2Q_4] \\
&\quad - \tfrac{1}{6}[-4T_2 + 40 + 2(-4T_3 + 40) + 2(-4T_3 + 40) - 4T_4 + 40]
\end{aligned} \qquad (2.4.34)$$

Inserting the values $T_1 = 10$, $T_2 = 10$, $T_4 = 10$, specified by the boundary conditions, and the values $Q_1 = 0$, $Q_2 = 4$, $Q_3 = 8$, $Q_4 = 4$, specified by the problem definition, into this equation, leads to the result $T_3 = 9$. The integrated nodal heat fluxes q_1, q_2 and q_4 can then be obtained by employing the equations corresponding to nodes 1, 2 and 4 respectively.

2.4.2   The use of numerical integration

In a computational implementation, the entries in the element stiffness and mass matrices of equations (2.4.13) and (2.4.14) can be conveniently calculated using numerical integration techniques. To see how this can be accomplished, we first observe that, in the current context, we are faced with the evaluation of integrals of the form

$$\int_e g(x, y)\,dx\,dy = \int_{-1}^{1} \int_{-1}^{1} G(\xi, \eta)\,d\xi\,d\eta \qquad (2.4.35)$$

where G is a polynomial in ξ and η. A straightforward approach which can be adopted here is to use the Gauss–Legendre numerical integration formulae, introduced in Section 2.2.3, to perform separate numerical integrations in the ξ and η directions. For example, using the formula of equation (2.2.34), the inner integral may be evaluated to give

$$\int_{-1}^{1} G(\xi, \eta)\,d\xi \approx \sum_{j=1}^{n_s} \omega_j G(\xi_j, \eta) \qquad (2.4.36)$$

This can be followed by a similar integration in the other direction, so that

$$\int_e g(x, y)\,dx\,dy \approx \sum_{j=1}^{n_s} \omega_j \left[\int_{-1}^{1} G(\xi_j, \eta)\,d\eta \right] \approx \sum_{j=1}^{n_s} \omega_j \left[\sum_{k=1}^{n_s} \omega_k G(\xi_j, \eta_k) \right] \qquad (2.4.37)$$

resulting finally in the approximation

$$\int_e g(x, y)\,dx\,dy \approx \sum_{j=1}^{n_s} \sum_{k=1}^{n_s} \omega_{jk} G(\xi_j, \eta_k) \qquad \omega_{jk} = \omega_j \omega_k \qquad (2.4.38)$$

For this formula, the sampling points are located at (ξ_j, η_k) and the exact location of the points along with the specification of the corresponding weights, will be determined when the value of n_s is selected. Note that, if the integration procedure is exact for polynomials of degree p, in the ξ and η directions separately, then the formula of equation (2.4.38) will integrate exactly all terms of the form $\xi^{p_1}\eta^{p_2}$ whenever $p_1, p_2 \leqslant p$. From the observations made in Section 2.2.3, it follows that the choice $n_s = 2$ will be necessary, if this approach is used to evaluate exactly the entries in the rectangular element mass and stiffness matrices of equations (2.4.13) and (2.4.14).

The basic Gauss–Legendre method of Section 2.2.3 can be used to evaluate the entries in the side mass matrix of equation (2.4.29). Exact evaluation will follow when the value $n_s = 2$ is adopted in this case also.

2.4.3   The general four-noded isoparametric element

Now we consider the solution of the heat conduction problem defined in equations (2.4.1)–(2.4.3) when the domain Ω is of more general shape, as shown in Figure 2.4.4(a). Four nodes, numbered 1, 2, 3 and 4, are located on the boundary Γ of Ω, such that, in this case, nodes 2 and 4 are located at the points where the segments Γ_1 and Γ_2 meet. Then, the standard four-noded square element of Section 1.6.1, defined on the range $-1 \leqslant \xi, \eta \leqslant 1$, can be mapped into an element e which approximates the domain Ω, as shown in Figure 2.4.4(b). The sides of the element e are numbered from 1 to 4, as shown in Figure 2.4.4(b). The element shape functions are as defined in equation (2.4.4), and the isoparametric mapping is defined by

$$x = \sum_{j=1}^{4} x_j N_j(x, y) = \alpha_{1e} + \alpha_{2e}\xi + \alpha_{3e}\eta + \alpha_{4e}\xi\eta$$

$$y = \sum_{j=1}^{4} y_j N_j(x, y) = \beta_{1e} + \beta_{2e}\xi + \beta_{3e}\eta + \beta_{4e}\xi\eta$$

(2.4.39)

where $\alpha_{1e}, \ldots, \alpha_{4e}, \beta_{1e}, \ldots, \beta_{4e}$ are constants which depend on the geometry of the element e. In general, these constants will be non-zero, so that both x and y will be defined in terms of expressions which are bilinear in ξ and η. Working with trial functions

$$\hat{T} = \sum_{j=1}^{4} T_j N_j(\xi, \eta)$$

(2.4.40)

the finite element approximate solution is again determined from the equations

$$\sum_{j=1}^{4}\left[\int_e k\left(\frac{\partial N_j}{\partial x}\frac{\partial N_i}{\partial x} + \frac{\partial N_j}{\partial y}\frac{\partial N_i}{\partial y}\right)\mathrm{d}x\,\mathrm{d}y\right]T_j = \sum_{j=1}^{4}\left[\int_e N_j N_i\,\mathrm{d}x\,\mathrm{d}y\right]Q_j - q_i$$

$$i = 1, 2, 3, 4 \qquad (2.4.41)$$

To perform the integration over the square ξ, η element, we use the mapping

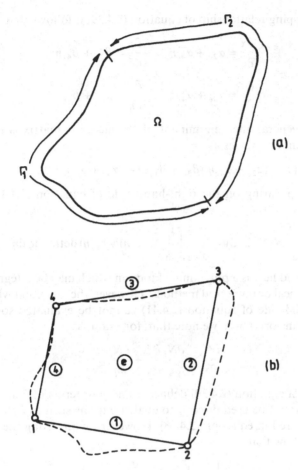

Figure 2.4.4 *Use of a four-noded quadrilateral element to approximate a general domain* Ω.

of equation (2.4.39). For a general mapping transformation, an element of area $dx\,dy$ is transformed according to

$$dx\,dy = \det(\mathbf{J})\,d\xi\,d\eta \qquad (2.4.42)$$

where $\mathbf{J}$ is the jacobian matrix of the transformation which is defined by

$$\mathbf{J} = \begin{bmatrix} \partial x/\partial \xi & \partial y/\partial \xi \\ \partial x/\partial \eta & \partial y/\partial \eta \end{bmatrix} \qquad (2.4.43)$$

so that

$$\det(\mathbf{J}) = \frac{\partial x}{\partial \xi}\frac{\partial y}{\partial \eta} - \frac{\partial x}{\partial \eta}\frac{\partial y}{\partial \xi} \qquad (2.4.44)$$

Using the mapping relationship of equation (2.4.39) it follows that

$$\frac{\partial x}{\partial \xi} = \alpha_{2e} + \alpha_{4e}\eta \qquad \frac{\partial y}{\partial \xi} = \beta_{2e} + \beta_{4e}\eta$$

$$\frac{\partial x}{\partial \eta} = \alpha_{3e} + \alpha_{4e}\xi \qquad \frac{\partial y}{\partial \eta} = \beta_{3e} + \beta_{4e}\xi \qquad (2.4.45)$$

Therefore, in general, the determinant of the jacobian matrix is not constant over the element and varies as

$$\det(\mathbf{J}) = (\alpha_{2e} + \alpha_{4e}\eta)(\beta_{3e} + \beta_{4e}\xi) - (\alpha_{3e} + \alpha_{4e}\eta)(\beta_{2e} + \beta_{4e}\xi) \qquad (2.4.46)$$

The integral appearing on the right-hand side of equation (2.4.41) is readily transformed as

$$\int_e N_j N_i \, dx \, dy = \int_{-1}^{1} \int_{-1}^{1} N_j(\xi, \eta) N_i(\xi, \eta) \det(\mathbf{J}) \, d\xi \, d\eta \qquad (2.4.47)$$

and the integrand here is a polynomial function which may be integrated exactly by either analytical or numerical methods. However, the integrand which appears on the left-hand side of equation (2.4.41) cannot be evaluated so readily. To perform the transformation, we note that, for example,

$$\frac{\partial N_i}{\partial x} = \frac{\partial N_i}{\partial \xi} \frac{\partial \xi}{\partial x} + \frac{\partial N_i}{\partial \eta} \frac{\partial \eta}{\partial x} \qquad (2.4.48)$$

The mapping of equation (2.4.39) defines x and y in terms of ξ and η, and this information cannot be used directly to evaluate terms such as $\partial \xi / \partial x$ and $\partial \eta / \partial x$ which are required in equation (2.4.48). However, by inverting the mapping, it is possible to show that

$$\frac{\partial \xi}{\partial x} = \frac{1}{\det(\mathbf{J})} \frac{\partial y}{\partial \eta} \qquad \frac{\partial \xi}{\partial y} = -\frac{1}{\det(\mathbf{J})} \frac{\partial x}{\partial \eta}$$

$$\frac{\partial \eta}{\partial x} = -\frac{1}{\det(\mathbf{J})} \frac{\partial y}{\partial \xi} \qquad \frac{\partial \eta}{\partial y} = \frac{1}{\det(\mathbf{J})} \frac{\partial x}{\partial \xi} \qquad (2.4.49)$$

and, using the results of equations (2.4.42), (2.4.48) and (2.4.49), we see that

$$\int_e k \left(\frac{\partial N_j}{\partial x} \frac{\partial N_i}{\partial x} + \frac{\partial N_j}{\partial y} \frac{\partial N_i}{\partial y} \right) dx \, dy$$

$$= \int_{-1}^{1} \int_{-1}^{1} \frac{k}{\det(\mathbf{J})} \left(\frac{\partial N_j}{\partial \xi} \frac{\partial y}{\partial \eta} - \frac{\partial N_j}{\partial \eta} \frac{\partial y}{\partial \xi} \right) \left(\frac{\partial N_i}{\partial \xi} \frac{\partial y}{\partial \eta} - \frac{\partial N_i}{\partial \eta} \frac{\partial y}{\partial \xi} \right) d\xi \, d\eta$$

$$+ \int_{-1}^{1} \int_{-1}^{1} \frac{k}{\det(\mathbf{J})} \left(-\frac{\partial N_j}{\partial \xi} \frac{\partial x}{\partial \eta} + \frac{\partial N_j}{\partial \eta} \frac{\partial x}{\partial \xi} \right) \left(-\frac{\partial N_i}{\partial \xi} \frac{\partial x}{\partial \eta} + \frac{\partial N_i}{\partial \eta} \frac{\partial x}{\partial \xi} \right) d\xi \, d\eta \qquad (2.4.50)$$

The evaluation of integrals of this form becomes difficult, if not impossible in general, by analytical methods, but may be accomplished approximately by using the numerical integration procedures of Section 2.4.2.

When the entries in the 4×4 element mass and stiffness matrices have been evaluated, the equation set (2.4.41) may be expressed in the matrix form

$$\mathbf{K}_e \mathbf{T}_e = \mathbf{M}_e \mathbf{Q}_e - \mathbf{q}_e \tag{2.4.51}$$

where the element vectors $\mathbf{T}_e, \mathbf{Q}_e$ and $\mathbf{q}_e$ are as defined in equation (2.4.15).

We illustrate the practical use of the general four-noded element by considering a specific example.

Example 2.6

Consider the problem of determining the distribution of temperature in a region Ω which is in the shape of a quadrant of a circular annulus, of inner radius $r = 1$ and outer radius $r = 2$, as shown in Figure 2.4.5(a). The origin of the (x, y) coordinate system is located at the centre of the circles forming the annulus, so that $r^2 = x^2 + y^2$. The region is made of material of constant thermal conductivity $k = 2$ and heat is being generated within the material at a rate $Q = 4(x^2 + y^2)$ per unit area. On the boundary $r = 1$, the temperature is maintained at the value $T = 10 - x^2 + x^4$, while the temperature is maintained at the value $T = 10$ on the boundary $y = 0$. The value of the heat flux is specified on the remaining boundaries, with $q = \aleph = 4x^2y^2$ on $r = 2$ and $q = \aleph = 0$ on $x = 0$.

Four nodes are numbered and located as shown in Figure 2.4.5(b) and the region of interest is then approximated by the element e which is formed by an isoparametric mapping from the standard four-noded square element. The problem specification implies that $T_1 = T_2 = T_4 = 10$, so that the unknowns in this case are T_3, q_1, q_2 and q_4.

Using numerical integration, with the choice $n_s = 2$ employed in both directions, the element stiffness and mass matrices are evaluated as

$$\mathbf{M}_e = \begin{bmatrix} 0.138\,889 & 8.333\,33 \times 10^{-2} & 4.166\,67 \times 10^{-2} & 6.944\,44 \times 10^{-2} \\ 8.333\,33 \times 10^{-2} & 0.194\,44 & 9.722\,22 \times 10^{-2} & 4.166\,67 \times 10^{-2} \\ 4.166\,67 \times 10^{-2} & 9.722\,22 \times 10^{-2} & 0.194\,444 & 8.333\,33 \times 10^{-2} \\ 6.944\,44 \times 10^{-2} & 4.166\,67 \times 10^{-2} & 8.333\,33 \times 10^{-2} & 0.138\,89 \end{bmatrix}$$

$$\tag{2.4.52a}$$

$$\mathbf{K}_e = \begin{bmatrix} 4.230\,77 & -4.615\,38 & -0.384\,62 & 0.769\,23 \\ -4.615\,38 & 7.307\,69 & 2.692\,31 & -5.8462 \\ -0.384\,62 & 2.692\,31 & 2.307\,69 & -4.615\,38 \\ 0.769\,23 & -5.384\,62 & -4.615\,38 & 9.230\,77 \end{bmatrix} \tag{2.4.52b}$$

With the sides of the element e numbered as shown in Figure 2.4.5(b), the local

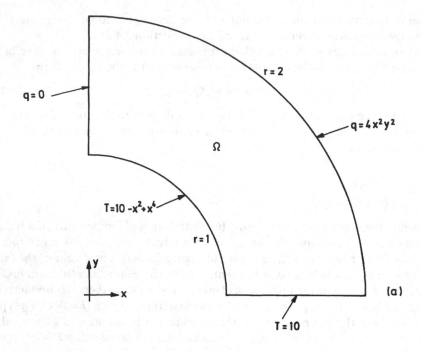

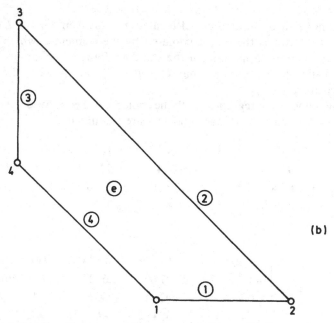

Figure 2.4.5 *(a) Problem domain and boundary conditions. (b) Four-noded element used in constructing an approximate solution.*

and global node numbers for each side are related according to

side	local node 1	local node 2	
1	1	2	
2	2	3	(2.4.53a)
3	3	4	
4	4	1	

and then

$$q_1 = \int_4 \hat{q} N_1 \, d\Gamma + \int_1 \hat{q} N_1 \, d\Gamma \qquad q_2 = \int_1 \hat{q} N_2 \, d\Gamma + \int_2 \aleph N_2 \, d\Gamma$$

$$\qquad\qquad\qquad\qquad\qquad\qquad\qquad\qquad\qquad\qquad (2.4.53b)$$

$$q_3 = \int_2 \aleph N_3 \, d\Gamma + \int_3 \aleph N_3 \, d\Gamma \qquad q_4 = \int_3 \aleph N_4 \, d\Gamma + \int_4 \hat{q} N_4 \, d\Gamma$$

These integrals may now be evaluated using the integration rules of equations (2.4.30) and, in this way, we can deduce that

$$q_3 = 0 \qquad\qquad\qquad (2.4.54)$$

since $\aleph_2 = \aleph_3 = \aleph_4 = 0$ for this example. The equation corresponding to node 3 is then obtained from equation (2.4.51) as

$$-0.384\,62 T_1 + 2.692\,31 T_2 + 2.307\,69 T_3 - 4.615\,38 T_4$$
$$= 4.166\,67 \times 10^{-2} Q_1 + 9.722\,22 \times 10^{-2} Q_2 + 0.194\,444 Q_3 + 8.333\,33 \times 10^{-2} Q_4$$

where, by definition, $Q_1 = Q_4 = 4$ and $Q_2 = Q_3 = 16$. Inserting the known values, we deduce that $T_3 = 12.08$. The integrated nodal heat fluxes q_1, q_2 and q_4 may then be obtained by solving the equations corresponding to nodes 1, 2 and 4 respectively.

2.4.4   The validity of the mapping

Since both (x, y) and (ξ, η) have been defined here as right-handed coordinate systems, it can be seen from equation (2.4.42) that the isoparametric mapping will be a proper mapping between these two coordinate systems provided that $\det(\mathbf{J}) > 0$ at all points. The presence of points at which $\det(\mathbf{J}) = 0$ or $\det(\mathbf{J}) < 0$ will indicate that the mapping is not invertible, as can be seen from equation (2.4.49). For this element, it can be shown that the mapping employed in equation (2.4.39) will be invertible provided all angles of the element e are less than $180°$.

2.4.5   The eight-noded rectangular element

Consider again the solution of the problem defined via equations (2.4.1)–(2.4.3). When the solution domain Ω is the rectangle considered in Section 2.4.1, a

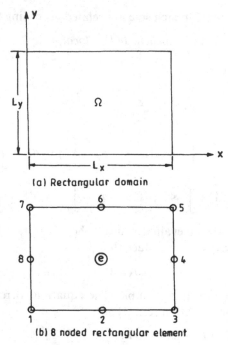

(a) Rectangular domain

(b) 8 noded rectangular element

Figure 2.4.6 *Rectangular domain and eight-noded rectangular element.*

solution of improved overall accuracy can be produced by employing a higher-order element. To illustrate this process, we locate eight nodes, numbered from 1 to 8, on the boundary Γ of Ω, as shown in Figure 2.4.6, with nodes 2, 4, 6 and 8 being centrally located on their respective sides. The standard eight-noded square element of Section 1.6.3, defined on the range $-1 \leqslant \xi, \eta \leqslant 1$, can be mapped into an element e which exactly represents the domain Ω. This mapping is accomplished by using the element shape functions of equation (1.6.20), viz

$$
\begin{aligned}
N_1 &= (1 - \xi)(1 - \eta)(-1 - \xi - \eta)/4 & N_2 &= (1 - \xi^2)(1 - \eta)/2 \\
N_3 &= (1 + \xi)(1 - \eta)(-1 + \xi - \eta)/4 & N_4 &= (1 + \xi)(1 - \eta^2)/2 \\
N_5 &= (1 + \xi)(1 + \eta)(-1 + \xi + \eta)/4 & N_6 &= (1 - \xi^2)(1 + \eta)/2 \\
N_7 &= (1 - \xi)(1 + \eta)(-1 - \xi + \eta)/4 & N_8 &= (1 - \xi)(1 - \eta^2)/2
\end{aligned}
\tag{2.4.55}
$$

In this case, the isoparametric relation is expressed as

$$
x = \sum_{j=1}^{8} x_j N_j(\xi, \eta) = h_{xe}/2 + h_{xe}\xi/2
$$

$$
y = \sum_{j=1}^{8} y_j N_j(\xi, \eta) = h_{ye}/2 + h_{ye}\eta/2
$$

$$\tag{2.4.56}$$

since

$$x_1 = x_7 = x_8 = 0 \qquad x_2 = x_6 = h_{xe}/2 \qquad x_3 = x_4 = x_5 = h_{xe}$$
$$y_1 = y_2 = y_3 = 0 \qquad y_4 = y_8 = h_{ye}/2 \qquad y_5 = y_6 = y_7 = h_{ye} \qquad (2.4.57)$$

It can be observed that the mapping here is exactly the same as that employed with the four-noded rectangular element in equation (2.4.5). Again, following the approach of Section 1.7, we work with trial functions

$$\hat{T} = \sum_{j=1}^{8} T_j N_j(\xi, \eta) \qquad (2.4.58)$$

and the standard finite element procedure leads to the equation system

$$\mathbf{K}_e \mathbf{T}_e = \mathbf{M}_e \mathbf{Q}_e - \mathbf{q}_e \qquad (2.4.59)$$

where the 8×8 matrices $\mathbf{K}_e$ and $\mathbf{M}_e$ are the eight-noded rectangular element stiffness and mass matrices respectively, with typical entries

$$[\mathbf{K}_e]_{ij} = \frac{1}{h_{xe}h_{ye}} \sum_{j=1}^{8} \left[\int_{-1}^{1} \int_{-1}^{1} k \left(h_{ye}^2 \frac{\partial N_j}{\partial \xi} \frac{\partial N_i}{\partial \xi} + h_{xe}^2 \frac{\partial N_j}{\partial \eta} \frac{\partial N_i}{\partial \eta} \right) d\xi \, d\eta \right] \qquad (2.4.60)$$

and

$$[\mathbf{M}_e]_{ij} = \frac{h_{xe}h_{ye}}{4} \sum_{j=1}^{8} \left[\int_{-1}^{1} \int_{-1}^{1} N_j N_i \, d\xi \, d\eta \right] \qquad (2.4.61)$$

In this case, the vectors $\mathbf{T}_e, \mathbf{Q}_e$ and $\mathbf{q}_e$ are given by

$$\mathbf{T}_e = \begin{bmatrix} T_1 \\ T_2 \\ \vdots \\ T_8 \end{bmatrix} \qquad \mathbf{Q}_e = \begin{bmatrix} Q_1 \\ Q_2 \\ \vdots \\ Q_8 \end{bmatrix} \qquad \mathbf{q}_e = \begin{bmatrix} q_1 \\ q_2 \\ \vdots \\ q_8 \end{bmatrix} \qquad (2.4.62)$$

As in the previous examples which we have considered, the entries in the element stiffness and mass matrices may be evaluated analytically, but this becomes a less attractive prospect as the polynomial order and the size of the matrices increase. The entries here are therefore best evaluated by numerical integration techniques, with exact evaluation resulting from the use of an appropriate number of sampling points. The highest-order terms which need to be integrated when determining the entries in the mass or stiffness matrices of equations (2.4.60) and (2.4.61) are of the form $\xi^4\eta^2$ and $\eta^4\xi^2$ and hence all the entries in these matrices can be evaluated exactly be employing the integration rule of equation (2.4.37) with the choice $n_s = 3$ and a total of nine sampling points.

Example 2.7

To illustrate the imposition of boundary conditions for this element, we consider again the problem outlined in Example 2.5. In this case, the problem is described in Figure 2.4.7, where the node and side numbering which has been employed

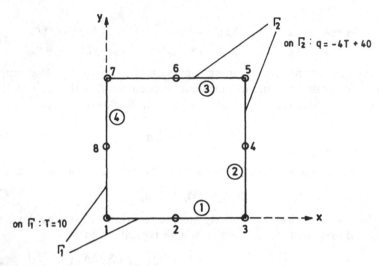

Figure 2.4.7 *Node numbering and boundary conditions for the eight-noded elements used in Example 2.7.*

is apparent. The entries in the element mass and stiffness matrices are evaluated and the element equation system (2.4.59) is formulated. The local and global node numbers for each side are related according to

side	local node 1	local node 2	local node 3	
1	1	2	3	
2	3	4	5	(2.4.63a)
3	5	6	7	
4	7	8	1	

and then

$$q_1 = \int_4 \hat{q} N_1 \, d\Gamma + \int_1 \hat{q} N_1 \, d\Gamma \qquad q_4 = \int_2 \aleph N_4 \, d\Gamma$$

$$q_2 = \int_1 \hat{q} N_2 \, d\Gamma \qquad q_5 = \int_2 \aleph N_5 \, d\Gamma + \int_3 \aleph N_5 \, d\Gamma$$

$$q_8 = \int_4 \hat{q} N_8 \, d\Gamma \qquad q_6 = \int_3 \aleph N_6 \, d\Gamma \qquad (2.4.63b)$$

$$q_3 = \int_1 \hat{q} N_3 \, d\Gamma + \int_2 \aleph N_3 \, d\Gamma$$

$$q_7 = \int_3 \aleph N_7 \, d\Gamma + \int_4 \hat{q} N_7 \, d\Gamma$$

Again, the evaluation of these line integrals is conveniently accomplished by mapping followed by the use of analytical or numerical techniques. Since the element sides here are of unit length and with $\aleph$ approximated by its quadratic interpolant on both side 2 and side 3, the approach developed in Example 2.5, using the results of equation (2.2.30), can be followed and analytical integration employed to produce the expressions

$$q_4 = \tfrac{1}{30}[2\aleph_3 + 16\aleph_4 + 2\aleph_5]$$

$$q_5 = \tfrac{1}{30}[-\aleph_3 + 2\aleph_4 + 4\aleph_5] + \tfrac{1}{30}[4\aleph_5 + 2\aleph_6 - \aleph_7] \qquad (2.5.64)$$

$$q_6 = \tfrac{1}{30}(2\aleph_5 + 16\aleph_6 + 2\aleph_7)$$

The problem boundary conditions define the values $T_i = 10$, for $i = 1, 2, 3, 7, 8$, and $\aleph_i = -4T_i + 40$, for $i = 3, 4, 5, 6, 7$. Now the equations corresponding to nodes 4, 5 and 6 can be completely formulated and solved to produce values for T_4, T_5 and T_6. The remaining equations can be used, if required, to determine the integrated heat fluxes at nodes 1, 2, 3, 7 and 8.

2.4.6 The general eight-noded isoparametric element

We now consider the use of the eight-noded element to produce an approximate solution of the heat conduction problem defined in equations (2.4.1)–(2.4.3), when the domain Ω is of general shape. Eight nodes, numbered 1 to 8, are located on the boundary Γ of Ω such that nodes 3 and 7 are placed at the points where the segments Γ_1 and Γ_2 meet. An isoparametric mapping is employed to map the standard eight-noded element of Section 1.6.3, defined on the range $-1 \leqslant \xi, \eta \leqslant 1$, into an element e which approximates the domain Ω, as shown in Figure 2.4.8. With the element shape functions given in equation (2.4.55), the isoparametric mapping is defined by

$$x = \sum_{j=1}^{8} x_j N_j(\xi, \eta) = \alpha_{1e} + \alpha_{2e}\xi + \alpha_{3e}\eta + \alpha_{4e}\xi^2 + \alpha_{5e}\xi\eta + \alpha_{6e}\eta^2 + \alpha_{7e}\xi^2\eta + \alpha_{8e}\xi\eta^2$$

$$y = \sum_{j=1}^{8} y_j N_j(\xi, \eta) = \beta_{1e} + \beta_{2e}\xi + \beta_{3e}\eta + \beta_{4e}\xi^2 + \beta_{5e}\xi\eta + \beta_{6e}\eta^2 + \beta_{7e}\xi^2\eta + \beta_{8e}\xi\eta^2$$

$$(2.4.65)$$

where $\alpha_{1e}, \alpha_{2e}, \ldots, \alpha_{8e}, \beta_{1e}, \beta_{2e}, \ldots, \beta_{8e}$ are constants which depend upon the geometry of the element e. Working with trial functions

$$\hat{T} = \sum_{j=1}^{8} T_j N_j(\xi, \eta) \qquad (2.4.66)$$

the Galerkin procedure again leads to the equation system

$$\mathbf{K}_e \mathbf{T}_e = \mathbf{M}_e \mathbf{Q}_e - \mathbf{q}_e \qquad (2.4.67)$$

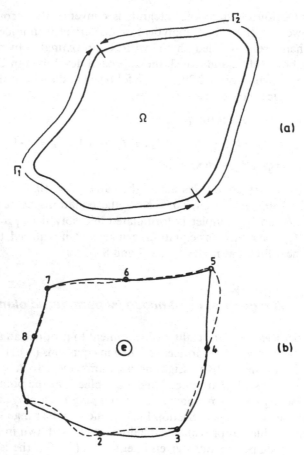

Figure 2.4.8 *Use of an eight-noded quadrilateral element e to approximate a general domain Ω.*

The results of Section 2.4.3 imply that, in this case, the entries in the 8×8 element stiffness and mass matrices may be computed as

$$[\mathbf{K}_e]_{ij} = \int_{-1}^{1} \int_{-1}^{1} \frac{k}{\det(\mathbf{J})} \left(\frac{\partial N_j}{\partial \xi} \frac{\partial y}{\partial \eta} - \frac{\partial N_j}{\partial \eta} \frac{\partial y}{\partial \xi} \right) \left(\frac{\partial N_i}{\partial \xi} \frac{\partial y}{\partial \eta} - \frac{\partial N_i}{\partial \eta} \frac{\partial y}{\partial \xi} \right) d\xi \, d\eta$$

$$+ \int_{-1}^{1} \int_{-1}^{1} \frac{k}{\det(\mathbf{J})} \left(-\frac{\partial N_j}{\partial \xi} \frac{\partial x}{\partial \eta} + \frac{\partial N_j}{\partial \eta} \frac{\partial x}{\partial \xi} \right) \left(-\frac{\partial N_i}{\partial \xi} \frac{\partial x}{\partial \eta} + \frac{\partial N_i}{\partial \eta} \frac{\partial x}{\partial \xi} \right) d\xi \, d\eta$$

$$(2.4.68)$$

and

$$[\mathbf{M}_e]_{ij} = \int_{-1}^{1} \int_{-1}^{1} N_j(\xi, \eta) N_i(\xi, \eta) \det(\mathbf{J}) \, d\xi \, d\eta \qquad (2.4.69)$$

The general expression for the jacobian matrix $\mathbf{J}$ is given in equation (2.4.43) which means, using equation (2.4.65), that for this example

$$\det(\mathbf{J}) = (\alpha_{2e} + 2\alpha_{4e}\xi + \alpha_{5e}\eta + 2\alpha_{7e}\xi\eta + \alpha_{8e}\eta^2)(\beta_{3e} + \beta_{5e}\xi + 2\beta_{6e}\eta + \beta_{7e}\xi^2 + 2\beta_{8e}\xi\eta)$$
$$- (\alpha_{3e} + \alpha_{5e}\xi + 2\alpha_{6e}\eta + \alpha_{7e}\xi^2 + 2\alpha_{8e}\xi\eta)(\beta_{2e}\xi + 2\beta_{4e}\xi + \beta_{5e}\eta + 2\beta_{7e}\xi\eta + \beta_{8e}\eta^2)$$

$$(2.4.70)$$

Numerical integration is employed to evaluate the entries in the stiffness and mass matrices in this case.

We illustrate the practical use of the general eight-noded element by considering a specific example.

Example 2.8

Consider again the problem of Example 2.6, which requires the computation of the distribution of temperature in a region which is in the shape of a quadrant of a circular annulus, of inner radius $r = 1$ and outer radius $r = 2$, as shown in Figure 2.4.5(a).

Figure 2.4.9 shows eight nodes located on the boundary of the region and the node and side numbering which is adopted. The nodes 2, 4, 6, 8 are centrally

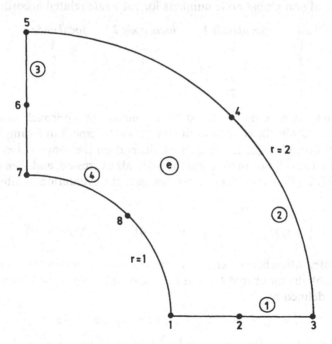

Figure 2.4.9 *Element and node numbers used in constructing an approximate solution to example 2.8.*

located on the respective boundary segments. The region of interest is then approximated by the element e which is formed by an isoparametric mapping from the standard eight-noded square element. The problem specification implies that $T_1 = T_2 = T_3 = T_7 = 10$, $T_8 = 9.75$, $\aleph_3 = \aleph_5 = \aleph_6 = \aleph_7 = 0$ and $\aleph_4 = 16$. The unknowns in the finite element formulation are $T_4, T_5, T_6, q_1, q_2, q_3, q_7$ and q_8.

Numerical integration, with the choice $n_s = 3$ employed in both directions, is used to evaluate the entries in the element stiffness and mass matrices of equation (2.4.67), while

$$q_1 = \int_4 \hat{q} N_1 \, d\Gamma + \int_1 \hat{q} N_1 \, d\Gamma \qquad q_2 = \int_1 \hat{q} N_2 \, d\Gamma$$

$$q_3 = \int_1 \hat{q} N_3 \, d\Gamma + \int_2 \aleph N_3 \, d\Gamma \qquad q_4 = \int_2 \aleph N_4 \, d\Gamma$$

$$q_5 = \int_2 \aleph N_5 \, d\Gamma + \int_3 \aleph N_5 \, d\Gamma \qquad q_6 = \int_3 \aleph N_6 \, d\Gamma \qquad (2.4.71)$$

$$q_7 = \int_3 \aleph N_7 \, d\Gamma + \int_4 \hat{q} N_7 \, d\Gamma \qquad q_8 = \int_4 \hat{q} N_8 \, d\Gamma$$

With the local and global node numbers for each side related according to

side	local node 1	local node 2	local node 3
1	1	2	3
2	3	4	5
3	5	6	7
4	7	8	1

(2.4.72a)

the line integrals may be evaluated by extending the approach proposed for similar quantities in the context of the four-noded element in Example 2.5. The standard three-noded quadratic element, defined on the range $-1 \leqslant \xi \leqslant 1$, can be mapped into each side in turn and the integrals expressed, and then evaluated, in terms of ξ. To illustrate this process, consider the evaluation of integrals such as

$$\int_s g(x, y) N_I \, d\Gamma \qquad \int_s g(x, y) N_J \, d\Gamma \qquad \int_s g(x, y) N_K \, d\Gamma \qquad (2.4.72b)$$

where the integration here extends over an element side s connecting three nodes, which are globally numbered I, J and K, as shown in Figure 2.4.10. The required mapping is defined by

$$x = x_{1s} N_1 + x_{2s} N_2 + x_{3s} N_3 = \alpha_{1s} + \alpha_{2s} \xi + \alpha_{3s} \xi^2$$
$$y = y_{1s} N_1 + y_{2s} N_2 + y_{3s} N_3 = \beta_{1s} + \beta_{2s} \xi + \beta_{3s} \xi^2 \qquad (2.4.73)$$

where $N_1(\xi)$, $N_2(\xi)$ and $N_3(\xi)$ are the standard quadratic element shape functions

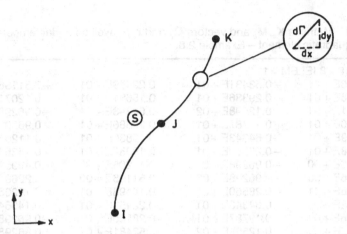

Figure 2.4.10 *Segments s of a boundary connecting three nodes I, J, K which form one side of a quadratic eight-noded element.*

defined in equation (2.2.22) and $x_{1s} = x_I$, $x_{2s} = x_J$, $x_{3s} = x_K$, $y_{1s} = y_I$, $y_{2s} = y_J$, $y_{3s} = y_K$. The coefficients $\alpha_{1s}, \alpha_{2s}, \alpha_{3s}, \beta_{1s}, \beta_{2s}$ ans β_{3s} will depend upon the co-ordinates of the nodes I, J and K. The element of length on the segment is defined according to equation (2.4.25) and may be computed for this element as

$$d\Gamma = \sqrt{(\alpha_{2s} + 2\alpha_{3s}\xi)^2 + (\beta_{2s} + 2\beta_{3s}\xi)^2}\, d\xi \qquad (2.4.74)$$

If g is quadratically interpolated, with respect to ξ, over the side s in terms of its nodal values then the required integrals can be expressed as

$$\int_s \begin{bmatrix} g(x,y)N_I \\ g(x,y)N_J \\ g(x,y)N_K \end{bmatrix} d\Gamma = \mathbf{M}_s \mathbf{g}_s \qquad (2.4.75a)$$

where $\mathbf{M}_s$ is the 3×3 side mass matrix, whose entries are defined by

$$[\mathbf{M}_s]_{ij} = \int_{-1}^{1} N_j(\xi)N_i(\xi)\sqrt{(\alpha_{2s} + 2\alpha_{3s}\xi)^2 + (\beta_{2s} + 2\beta_{3s}\xi)^2}\, d\xi \qquad (2.4.75b)$$

and the side vector $\mathbf{g}_s$ is given by

$$\mathbf{g}_s = \begin{bmatrix} g_{1s} \\ g_{2s} \\ g_{3s} \end{bmatrix} = \begin{bmatrix} g_I \\ g_J \\ g_K \end{bmatrix} \qquad (2.4.76)$$

Numerical integration may be applied to evaluate the entries in the side mass matrix of equation (2.4.75).

From the problem specification, it can be deduced that

$$\mathbf{Q}^T = [4, 9, 16, 16, 16, 9, 4, 4] \qquad (2.4.77)$$

Table 2.1 Matrices K_e, M_e and vectors Q_e and r_e as well as nodal temperatures for a single quadratic element—Example 2.8.

K_e matrix 8 × 8 IELEM = 1

0.42560E + 01	−0.53481E + 01	0.22459E + 01	−0.51196E + 00
0.19358E + 01	−0.29386E + 01	0.15681E + 01	−0.12071E + 01
−0.53481E + 01	0.12598E + 02	−0.64343E + 01	−0.96298E + 00
−0.28560E + 01	0.49787E + 01	−0.29386E + 01	0.96298E + 00
0.22459E + 01	−0.64343E + 01	0.49832E + 01	0.11205E + 00
0.14498E + 01	−0.28560E + 01	0.19358E + 01	−0.14364E + 01
−0.51196E + 00	−0.96298E + 00	0.11205E + 00	0.49953E + 01
0.11205E + 00	−0.96298E + 00	−0.51196E + 00	−0.22695E + 01
0.19358E + 01	−0.28560E + 01	0.14498E + 01	0.11205E + 00
0.49832E + 01	−0.64343E + 01	0.22459E + 01	−0.14364E + 01
−0.29386E + 01	0.49787E + 01	−0.28560E + 01	−0.96298E + 00
−0.64343E + 01	0.12598E + 02	−0.53481E + 01	0.96298E + 00
0.15681E + 01	−0.29386E + 01	0.19358E + 01	−0.15196E + 00
0.22459E + 01	−0.53481E + 01	0.42560E + 01	−0.12071E + 01
−0.12071E + 01	0.96298E + 00	−0.14364E + 01	−0.22695E + 01
−0.14364E + 01	0.96298E + 00	−0.12071E + 01	0.56307E + 01

M_e matrix 8 × 8 IELEM = 1

0.68376E − 01	−0.81328E − 01	0.22189E − 01	−0.10536E + 00
0.36045E − 01	−0.99803E − 01	0.26808E − 01	−0.70250E − 01
−0.81328E − 01	0.42130E + 01	−0.62853E − 01	0.26614E + 00
−0.99803E − 01	0.19961E + 01	−0.99803E − 01	0.23287E + 00
0.22189E − 01	−0.62853E − 01	0.86851E − 01	−0.68409E − 01
0.17571E − 01	−0.99803E − 01	0.36045E − 01	−0.88724E − 01
−0.10536E + 00	0.26614E + 00	−0.68409E − 01	0.45286E + 00
−0.68409E − 01	0.26614E + 00	−0.10536E + 00	0.19408E + 00
0.36045E − 01	−0.99803E − 01	0.17571E − 01	−0.68409E − 01
0.86851E − 01	−0.62853E − 01	0.22189E − 01	−0.88724E − 01
−0.99803E − 01	0.19961E + 00	−0.99803E − 01	0.26614E + 00
−0.62853E − 01	0.42130E + 00	−0.81328E − 01	0.23287E + 00
0.26808E − 01	−0.99803E − 01	0.36045E − 01	−0.10536E + 00
0.22189E − 01	−0.81328E − 01	0.68376E − 01	−0.70250E − 01
−0.70250E − 01	0.23287E + 00	−0.88724E − 01	0.19408E + 00
−0.88724E − 01	0.23287E + 00	−0.70250E − 01	0.32347E + 00

Heat source vectors Q_e 8 × 1 IELEM = 1

0.40000E + 01	0.90000E + 01	0.16000E + 02	0.16000E + 02
0.16000E + 02	0.90000E + 01	0.40000E + 01	0.39999E + 01

RIGHT HAND VECTOR 8 × 1 IELEM = 1

−0.28099E + 00	0.93147E + 00	−0.35491E + 00	0.77632E + 00
−0.35489E + 00	0.93147E + 00	−0.28099E + 00	0.12939E + 01

The equations corresponding to nodes 4, 5, 6 are obtained from equation (2.4.67) and these equations solved to determine the nodal unknowns T_4, T_5, T_6. The integrated nodal heat fluxes q_1, q_2, q_3, q_7 and q_8 may be obtained directly, if required, from the equations corresponding to nodes 1, 2, 3, 7 and 8 respectively. Table 2.1 shows the matrices $\mathbf{K}_e$ and $\mathbf{M}_e$, vectors $\mathbf{Q}_e$ and $\mathbf{r}_e$ along with the nodal temperatures.

2.4.7 The validity of the mapping

It has been noted in Section 2.4.4, that an isoparametric mapping, of the form employed here, will be a proper mapping between two right-handed coordinate systems (x, y) and (ξ, η), provided that $\det(\mathbf{J}) > 0$ at all points. It can be shown that the mapping employed in equation (2.4.65) will be nonproper if, on any side, the distance between two nodes is less than one-third of the side length.

2.5 TWO-DIMENSIONAL PROBLEM SOLVED USING AN ASSEMBLY OF ELEMENTS

Earlier in this chapter, it was shown how solutions of improved accuracy could be obtained for one-dimensional heat conduction problems by subdividing the domain into a number of low-order elements. It was seen that, with the element matrices calculated for a general single element, the assembly of the global matrices for the whole domain could be accomplished in a straightforward manner. This process is directly extendable into two dimensions and is widely employed with the elements we have discussed here.

2.5.1 The use of four-noded rectangular elements

Consider again the solution of the heat conduction problem defined by equations (2.4.1)–(2.4.3), over the rectangular domain Ω defined by $0 \leqslant x \leqslant L_x, 0 \leqslant y \leqslant L_y$. In this case, the domain is subdivided to create a mesh of non-overlapping four-noded rectangular elements, as shown in Figure 2.5.1. As before, the nodes and the elements are numbered. Figure 2.5.2 shows a typical element e with nodes which are locally numbered 1, 2, 3, 4 and which are globally numbered I, J, K, L. If the element is of length h_{xe} in the x direction and of length h_{ye} in the y direction, the standard square bilinear element can be mapped exactly into this element e by the mapping

$$x = \sum_{j=1}^{4} x_{je} N_j(\xi, \eta) = x_I + h_{xe}/2 + h_{xe}\xi/2$$

$$y = \sum_{j=1}^{4} y_{je} N_j(\xi, \eta) = y_I + h_{ye}/2 + h_{ye}\eta/2$$

$$(2.5.1)$$

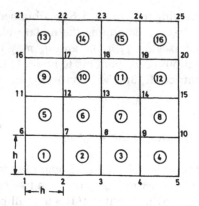

Figure 2.5.1 *Discretisation of a domain using four-noded rectangular elements.*

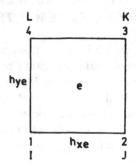

Figure 2.5.2 *A typical rectangular element e with local and global node numbers 1, 2, 3, 4, and I, J, K, L respectively.*

since

$$x_J = x_K = x_I + h_{xe} \qquad x_L = x_I$$
$$y_L = y_K = y_I + h_{ye} \qquad y_J = y_I \tag{2.5.2}$$

Over this element, we work with trial functions

$$\hat{T} = \sum_{j=1}^{4} T_{je} N_j(\xi, \eta) \tag{2.5.3}$$

and it is then apparent, from equations (2.4.7)–(2.4.11), that the approximation over the element is completed by the solution of the matrix equation system

$$\mathbf{K}_e \mathbf{T}_e = \mathbf{M}_e \mathbf{Q}_e - \mathbf{q}_e \tag{2.5.4}$$

Now the vectors $\mathbf{T}_e$, $\mathbf{Q}_e$ and $\mathbf{q}_e$ are given by

$$\mathbf{T}_e = \begin{bmatrix} T_{1e} \\ T_{2e} \\ T_{3e} \\ T_{4e} \end{bmatrix} \qquad \mathbf{Q}_e = \begin{bmatrix} Q_{1e} \\ Q_{2e} \\ Q_{3e} \\ Q_{4e} \end{bmatrix} \qquad \mathbf{q}_e = \begin{bmatrix} q_{1e} \\ q_{2e} \\ q_{3e} \\ q_{4e} \end{bmatrix} \qquad (2.5.5)$$

and the element stiffness and mass matrices are as defined in equations (2.4.16) and (2.4.17). The equations for each element in turn the obtained from this general equation set by relating the local node numbers to the appropriate global node numbers and employing the relevant element length and conductivity in each case. This process can be illustrated by demonstrating the formation of the equation corresponding to a typical interior node for the mesh shown in Figure 2.5.1. With the local and global node numbers for each element related according to

element	local node 1	local node 2	local node 3	local node 4
1	1	2	7	6
2	2	3	8	7
3	3	4	9	8
4	4	5	10	9
5	6	7	12	11
6	7	8	13	12
7	8	9	14	13
8	9	10	15	14
⋮	⋮	⋮	⋮	⋮

$$(2.5.6)$$

we select node 7 as a typical interior node and note that it belongs to the four elements numbered 1, 2, 5 and 6. We therefore need to examine the element equation (2.5.4) for each of these elements and identify the component corresponding to node 7 in each case. Assuming that $h_{xe} = h_{ye} = h$, the equations corresponding to node 7 can be seen to be

Element 1

$$\frac{k_1}{3}[-T_1 - T_2/2 + 2T_7 - T_6/2] = \frac{h^2}{36}[Q_1 + 2Q_2 + 4Q_7 + 2Q_6] - q_{7(1)} \quad (2.5.6a)$$

Element 2

$$\frac{k_2}{3}[-T_2/2 - T_3 - T_8/2 + 2T_7] = \frac{h^2}{36}[2Q_2 + Q_3 + 2Q_8 + 4Q_7] - q_{7(2)} \quad (2.5.6b)$$

Element 5

$$\frac{k_5}{3}[-T_6/2+2T_7-T_{12}/2-T_{11}]=\frac{h^2}{36}[2Q_6+4Q_7+2Q_{12}+Q_{11}]-q_{7(5)} \quad (2.5.6c)$$

Element 6

$$\frac{k_6}{3}[2T_7-T_8/2-T_{13}-T_{12}/2]=\frac{h^2}{36}[4Q_7+2Q_8+Q_{13}+2Q_{12}]-q_{7(6)} \quad (2.5.6d)$$

and the assembled equation for node 7 is obtained by adding these four equations together and using the continuity of flux requirement to set

$$q_{7(1)}+q_{7(2)}+q_{7(5)}+q_{7(6)}=0 \quad (2.5.7)$$

This equation is valid provided that there is no source or sink of heat at this node.

Node 15 can be regarded as a typical node on the boundary and it can be seen to belong to elements 8 and 12 only. Considering each of these elements in turn, we find that the equations corresponding to node 15 are

From element 8:

$$\frac{k_8}{3}[-T_9-T_{10}/2+2T_{15}-T_{14}/2]=\frac{1}{36}[Q_9+2Q_{10}+4Q_{15}+2Q_{14}]-q_{15(8)}$$

$$(2.5.8a)$$

From element 12:

$$\frac{k_{12}}{3}[-T_{14}/2+2T_{15}-T_{20}/2-T_{19}]=\frac{1}{36}[2Q_{14}+4Q_{15}+2Q_{20}+Q_{19}]-q_{15(12)}$$

$$(2.5.8b)$$

Adding these two equations produces the assembled equation for node 15. We note that in this case

$$q_{15(8)}+q_{15(12)}=\int_{(10-15)+(15-20)}\aleph N_{15}\,d\Gamma \quad (2.5.9)$$

2.5.2 The use of general eight-noded isoparametric elements

Consider again the solution of the heat conduction problem defined by equations (2.4.1)–(2.4.3), over the domain Ω which is shown in Figure 2.5.3. In this case, the domain is subdivided to create a mesh of non-overlapping eight-noded isoparametric elements. As before, the nodes and the elements are numbered. Figure 2.5.4 shows a typical element e with nodes which are locally numbered 1, 2, 3, 4, 5, 6, 7, 8 and which are globally numbered I, J, K, L, M, N, O, P. The element

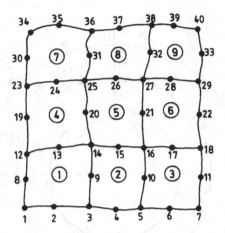

Figure 2.5.3 *Discretisation of a domain using eight-noded isoparametric elements.*

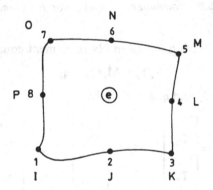

Figure 2.5.4 *A typical eight-noded isoparametric element with local and global node numbers.*

shape functions are given in equation (2.4.55) and the isoparametric mapping is defined by equation (2.4.65).

Over this element, the trial functions

$$\hat{T} = \sum_{j=1}^{8} T_{je} N_j(\xi, \eta) \qquad (2.5.10)$$

are utilised and it is then apparent, from equations (2.4.68)–(2.4.70), that the approximation over the elements is completed by the solution of the matrix

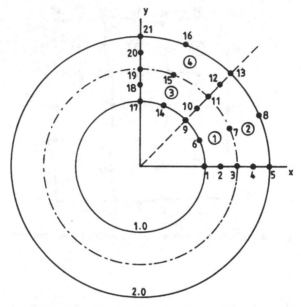

Figure 2.5.5 *Domain and discretisation·for Example 2.9.*

equation produced following the assembly of element equations of the form

$$\mathbf{K}_e \mathbf{T}_e = \mathbf{M}_e \mathbf{Q}_e - \mathbf{q}_e \qquad (2.5.11)$$

where the vectors $\mathbf{T}_e, \mathbf{Q}_e$ and $\mathbf{q}_e$ are given by

$$\mathbf{T}_e = \begin{bmatrix} T_{1e} \\ T_{2e} \\ \vdots \\ T_{8e} \end{bmatrix} \qquad \mathbf{Q}_e = \begin{bmatrix} Q_{1e} \\ Q_{2e} \\ \vdots \\ Q_{8e} \end{bmatrix} \qquad \mathbf{q}_e = \begin{bmatrix} q_{1e} \\ q_{2e} \\ \vdots \\ q_{8e} \end{bmatrix} \qquad (2.5.12)$$

Table 2.2   The comparison of the numerical solution with the exact solution

Node number	Analytical solution	One element	Two elements	Four elements
7	9.3672			8.4031
8	8.0000		7.1020	7.1395
10	9.3896			8.7950
11	8.7344		7.7074	7.7302
12	7.6553			6.7418
13	6.0000	0.0585	5.1168	5.5084

and the element stiffness and mass matrices are as defined in equations (2.4.68) and (2.4.69). The equations for each element in turn are obtained from the general equation set by relating the local node numbers to the appropriate global node numbers and employing the relevant element length and conductivity in each case. For this example, the local and global node numbers for each element are related according to

Element number	Local node numbers							
e	1	2	3	4	5	6	7	8
1	1	2	3	9	14	13	12	8
2	3	4	5	10	16	15	14	9
3	5	6	7	11	18	17	16	10
4	12	13	14	20	25	24	23	19
⋮								
9	27	28	29	33	40	39	38	32

We select node 14 as a typical interior node and note that it belongs to the four elements numbered 1, 2, 5 and 4. We therefore need to examine the element equation (2.5.11) for each of these elements and identify the component corresponding to node 14 in each case. The assembled equation for node 14 is obtained by adding four equations together and using the continuity of flux requirement to set

$$q_{14(1)} + q_{14(2)} + q_{14(5)} + q_{14(4)} = 0 \qquad (2.5.13)$$

This equation is valid provided that there is no source or sink of heat at this node.

We illustrate the practical use of an assembly of eight-noded isoparametric elements by considering a particular example.

Example 2.9

Consider again the problem of example 2.6, which requires the computation of the distribution of temperature in a region which is in the shape of a quadrant of a circular annulus, of inner radius $r = 1$ and outer radius $r = 2$ as shown in Figure 2.4.5(a).

Figure 2.5.5 shows the discretization of the domain by four eight-noded isoparametric elements. The nodes 2, 4, 10, 12, 18 and 20 are centrally located on the respective boundary segments. The problem specification implies that

$$T_1 = T_2 = T_3 = T_4 = T_5 = T_{21} = T_{20} = T_{19} = T_{18} = T_{17} = 10.0$$

$$T_9 = 9.75 \qquad T_{13} = 16 \qquad T_5 = T_4 = 0$$

The unknown temperatures in the finite element formulation are $T_6, T_7, T_8, T_{10}, T_{11}, T_{12}, T_{13}, T_{14}, T_{15}, T_{16}$.

This problem has been solved using one quadratic element, two quadratic elements and four quadratic elements. The numerical results produced in each case are compared with the exact solution for this problem in Table 2.2.

ADDITIONAL READING

Becker E. B., Carey G. F. and Oden J. T., *Finite Elements—An Introduction*, Volume 1, Prentice-Hall (1981).

Carlslaw H. S. and Jaeger J. C., *Conduction of Heat in Solids* (2nd Edition), Oxford University Press (1959).

Johnson C., *Numerical Solution of Partial Differential Equations by the Finite Element Method*, Cambridge University Press (1987).

Luikov A. V., *Analytical Heat Diffusion Theory*, Academic Press (1968).

Morris J. Ll., *Computational Methods in Elementary Numerical Analysis*, Academic Press (1968).

Reddy J. N., *An Introduction to the Finite Element Method*, Wiley-Interscience (1983).

Strang G. and Fix G. J., *An Analysis of the Finite Element Method*, Prentice-Hall (1973).

Zienkiewicz O. C. and Morgan K., *Finite Elements and Approximation*, John Wiley & Sons (1983).

Zienkiewicz O. C. and Taylor R. L., *The Finite Element Method* (4th Edition), Volume 1, McGraw-Hill (1989).

3 Time Stepping Methods for Heat Transfer

3.1 INTRODUCTION

In Chapter 2 we saw how the finite element method was used in the solution of the steady-state heat conduction problem. The solution of the differential equation was governed largely by known boundary conditions and, therefore, steady-state problems are often referred to as *boundary value problems*.

However, the majority of engineering problems are transient in nature and we are required to solve the time-dependent equation, which yields the solution at various times. In addition to the boundary conditions at any time, we are typically given the initial state of the system, at time $t = 0$. Transient problems can therefore be termed *initial value problems*. The transient heat conduction problem derived in Chapter 1 is given by equation (1.2.9)

$$\rho c \frac{\partial T}{\partial t} = \text{div}(k \, \text{grad} \, T) + Q \tag{3.1.1}$$

Application of the finite element spatial discretization in Section 1.3 finally yields the matrix equation system

$$\mathbf{M\dot{T}} + \mathbf{KT} = \mathbf{f} \tag{3.1.2}$$

where $\mathbf{M}$ is the capacitance matrix, $\mathbf{K}$ is the conductance matrix and $\mathbf{\dot{T}}$ is the temperature differentiated with respect to time.

Although an analytical solution of (3.1.2) is possible for simplified, linear cases, it is more usual for the system of ordinary differential equations to be discretised in time, from which solutions at various times can be obtained. Since the original partial differential equation becomes a discrete set of ordinary differential equations, the process is known as partial discretisation.

The remainder of this chapter is consequently concerned with time discretisation or time stepping schemes used in the solution of the parabolic first-order heat conduction equation (3.1.1) using, in general, two-level algorithms, where the solution at a particular time level is directly calculated from the temperatures at the previous time level.

Two techniques are employed to fully discretise the system of equations: the finite difference method and the finite element method. Consider first the finite difference method.

3.2   FINITE DIFFERENCE APPROXIMATIONS

Although finite difference schemes are special cases of the finite element method, the finite difference approximations are more widely used. In order to derive the various differencing schemes, consider the partial differential equation, which is a simplified version of (3.1.1), given by

$$\frac{\partial u}{\partial t} = \frac{\partial^2 u}{\partial x^2}$$

If we let $u_t = \partial u/\partial t$ and $u_{xx} = \partial^2 u/\partial x^2$, then we have

$$u_t = u_{xx} \tag{3.2.1}$$

In the finite element formulation, the spatial derivatives were implicit in the conductance matrix, $\mathbf{K}$. Therefore, the finite difference method is applied to both spatial and time discretisation of (3.2.1), which will also be helpful in presenting some accuracy and stability properties of the various schemes.

If we now wish to approximate the time derivative u_t in terms of values at discrete time intervals, Δt, such that

$$u^n \approx u(t^n) = u(n\Delta t)$$

we need to establish a rule whereby the next value of u is dependent upon time steps

$$u^{n+1} = g(u^n, u^{n-1}, \ldots, u^0)$$

The simplest representation of the derivative approximation is intuitively

$$u_t \approx \frac{u^{n+1} - u^n}{\Delta t} \tag{3.2.2}$$

and substitution into (3.2.1) leads to the simplest time stepping algorithm, the *forward difference method* or Euler's scheme, given by

$$u^{n+1} = u^n + \Delta t(u_{xx}^n) \tag{3.2.3}$$

This can be derived more formally from a Taylor's series expansion and omitting terms involving $(\Delta t)^2$ or higher:

$$u(t + \Delta t) = u(t) + \Delta t \cdot u_t(t) + \frac{(\Delta t)^2}{2} u_{tt}(t) + \cdots$$

$$u^{n+1} = u^n + \Delta t(u_t)^n \tag{3.2.4}$$

However, (3.2.1) gives $u_t = u_{xx}$ and so

$$u^{n+1} = u^n + \Delta t(u_{xx})^n$$

Equation (3.2.4) can also be written as

$$\frac{u(t + \Delta t) - u(t)}{\Delta t} = u_t(t) + \frac{\Delta t}{2} u_{tt}(t) + \cdots$$

Therefore, the error involved in the approximation to (3.2.3) is proportional to the time step length, Δt, or of the order Δt, $O(\Delta t)$. The forward difference scheme is therefore a first-order or $O(\Delta t)$ method. It can be seen from equation (3.2.3) that the value of u at the next time step, u^{n+1}, is directly evaluated from the current value, u^n. In the same way, a system of equations describing all grid points or nodal values of the variable leads to the direct solution of the matrix of unknown nodal values. The scheme is therefore *explicit*.

If the spatial derivative term, u_{xx}, is evaluated at the $n+1$ time level, we then have

$$\frac{u^{n+1} - u^n}{\Delta t} = u_{xx}^{n+1}$$

$$u^{n+1} - \Delta t u_{xx}^{n+1} = u^n \qquad (3.2.5)$$

or

$$u^{n+1} = u^n + \Delta t(u_t)^{n+1}$$

This is the *backward difference method* and the Taylor's series expansion would once again show that the scheme is first order, of $O(\Delta t)$. However, solution of (3.2.5) is not direct and the solution of a system of equations requires the inversion of a coefficient matrix, e.g. by Gaussian elimination. The scheme is therefore *implicit*.

The two schemes that have been described so far utilise the values of the variables at the current time step, u^n, to evaluate the value at the next step, u^{n+1}. Both schemes are therefore known as two-step schemes. A three-step, second-order ($O(\Delta t^2)$) approximation, known as the *central difference* scheme, can also be derived from the same principles. However, this algorithm can lead to severe difficulties, as will be shown later. For completeness the method is included here and is given by

$$u_t^n \approx \frac{u^{n+1} - u^{n-1}}{2\Delta t}$$

Having determined the three finite difference approximations for the temporal derivative, u_t, in equation (3.2.1), we must now introduce the spatial derivatives. If a one-dimensional domain extends from a to b, i.e. $[a, b]$, and is divided into N discrete intervals of length Δx such that

$$x_0 = a \qquad x_1 = a + \Delta x \qquad x_2 = a + 2\Delta x \ldots \qquad x_N = b$$

$$\Delta x = \frac{b - a}{N} \quad \text{and} \quad u_j^n \sim u(x_j, t^n) = u(a + j\Delta x, n\Delta t)$$

then the central difference approximation of u_{xx} is given by

$$u_{xx} \simeq \frac{\delta x^2 u_j^2}{(\Delta x)^2} = \frac{u_{j+1}^n - 2u_j^n + u_{j-1}^n}{(\Delta x)^2} \tag{3.2.6}$$

By once more utilising the Taylor's series expansion, this time in Δx, it can be shown that equation (3.2.6) is a second-order approximation to u_{xx}, $O(\Delta x^2)$. The central difference approximation may now be substituted into the three previous time stepping schemes to yield:

Forward difference (Euler, explicit)

$$u_j^{n+1} = u_j^n + \frac{\Delta t}{(\delta x)^2}[u_{j+1}^n - 2u_j^n + u_{j-1}^n] \tag{3.2.7}$$

with a truncation error $O(\Delta t, \Delta x^2)$

Backward difference (implicit)

$$u_j^{n+1} - \frac{\Delta t}{(\Delta x)^2}[u_{j+1}^{n+1} - 2u_j^{n+1} + u_{j-1}^{n+1}] = u_j^n \tag{3.2.8}$$

with a truncation error $O(\Delta t, \Delta x^2)$.

Central difference (three level, explicit)

$$u_j^{n+1} = u_j^{n-1} + \frac{2\Delta t}{(\Delta x^2)}[u_{j+1}^n - 2u_j^n + u_{j-1}^n] \tag{3.2.9}$$

with a truncation error $O(\Delta t^2, \Delta x^2)$.

Although the forward difference and backward difference methods provide explicit and implicit schemes, it is also possible to introduce a parameter θ that allows the time stepping scheme to vary between the two extremes. If $\theta = 0$, we have the explicit forward difference scheme, and conversely, if $\theta = 1$, the implicit backward difference. The θ-method is given by

$$\frac{u_j^{n+1} - u_j^n}{\Delta t} = \frac{1}{(\Delta x)^2}[\theta \delta^2 \times u_j^{n+1} + (1 - \theta)\delta^2 \times u_j^n] \tag{3.2.10}$$

where $0 < \theta < 1$ (N. B. θ and $(1 - \theta)$ may sometimes be reversed in other texts.)

In this case the truncation error (which refers to the omission of terms of higher order from a Taylor's series expansion) is $O(\Delta t, \Delta x^2)$, i.e. first order in time, second order in space. However, when $\theta = 1/2$ we have a second-order approximation in both time and space, i.e. $O(\Delta t^2, \Delta x^2)$. This method is said to use midpoint weighting and is known as the *Crank–Nicolson* method.

We have now obtained four time stepping algorithms with which to solve the simplified partial differential equation (3.2.1). Applying the same method to the transient heat conduction system of equations (3.1.2) we obtain:

Forward difference

$$\mathbf{M} \cdot \mathbf{T}^{n+1} = (\mathbf{M} - \Delta t \mathbf{K})\mathbf{T}^n + \Delta t \cdot \mathbf{f}^n \qquad (3.2.11)$$

Backward difference

$$(\mathbf{M} + \Delta t \mathbf{K})\mathbf{T}^{n+1} = \mathbf{M} \cdot \mathbf{T}^n + \Delta t \mathbf{f}^n \qquad (3.2.12)$$

Central difference

$$\mathbf{M} \cdot \mathbf{T}^{n+1} = \mathbf{M} \cdot \mathbf{T}^{n-1} - 2\Delta t \mathbf{K} \cdot \mathbf{T}^n + 2\Delta t \mathbf{f}^n \qquad (3.2.13)$$

θ-method

$$(\mathbf{M} + \theta \Delta t \mathbf{K})\mathbf{T}^{n+1} = (\mathbf{M} - (1 - \theta)\Delta t \mathbf{K})\mathbf{T}^n + \Delta t \mathbf{f}^n \qquad (3.2.14)$$

where $0 < \theta < 1$

In progressing from (3.2.8) to (3.2.10) the inherently explicit nature of the algorithm is lost to a certain degree. The finite element equation now requires inversion of the capacitance matrix, M, for solution. However, if M is diagonalised (or 'lumped'), then the solution requires the trivial inversion of a diagonal matrix, thereby returning the explicit nature of the algorithm.

Furthermore, treatment of the accuracy of the finite element equations is not quite so straightforward. Although the error analysis still applies to the time domain, such that forward and backward difference schemes are first-order accurate in time and Crank–Nicolson second-order accurate, an error analysis of the Galerkin spatial approximation reveals that local error estimates are in terms of the element dimensions. For example, if h represents the dimension of an element, then the spatial error is of the order h^2 as opposed to Δx^2, i.e. $O(h^2)$.

Example 3.1
To highlight the various merits and disadvantages of three of the four schemes detailed in this section, a solution will now be sought of equation (3.1.2) for the simple case of $\mathbf{K} = \mathbf{M} = 1, \mathbf{f} = 0$, solving for a single temperature in time, i.e.

$$\frac{\partial T}{\partial t} + T = 0$$

subject to the initial condition of $T(0) = 1.0$

The schemes, as given by equations (3.2.11), (3.2.12) and (3.2.14), are applied in the solution of the single equation with time steps $\Delta t = 0.2, 0.5, 0.8, 1.0, 2.0$ and 2.5. The results for each scheme are given in Figures 3.2.1, 3.2.2 and 3.2.3, from which it can be seen that for time steps up to 1.0, all schemes exhibit the correct form of behaviour with the Crank–Nicolson scheme yielding closer agreement between the different time steps.

However, when $\Delta t = 2.0$, the forward difference scheme oscillates between $+1$ and -1, and at $\Delta t = 2.5$ becomes totally divergent and is considered unstable. Although the Crank–Nicolson scheme shows a certain amount of oscillation with $\Delta t = 2.5$, the solution is convergent. The backward difference scheme, on the other hand, exhibits no oscillations or instabilities.

With these results in mind, we can now consider the stability of the schemes more formally. Essentially, the stability of a time stepping algorithm is the requirement that any errors that may be introduced in the approximation at any time step are damped out, or, at least, do not grow as fast as the true solution. The instability of an algorithm is characterised by non-physical oscillations in the numerical solution, which eventually increase uncontrollably in amplitude, swamping out the true solution and causing the solution scheme to break down.

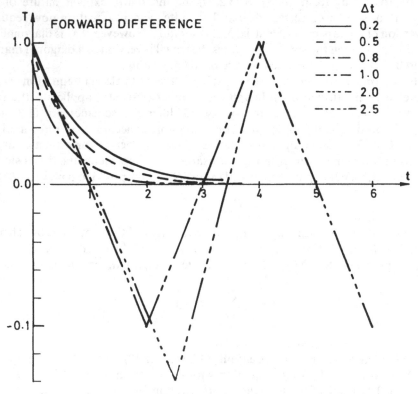

Figure 3.2.1   *Results of forward difference scheme.*

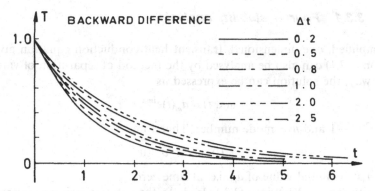

Figure 3.2.2   *Results of backward difference scheme.*

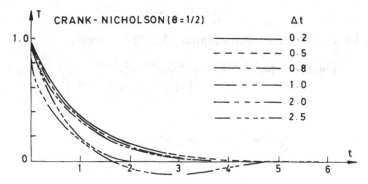

Figure 3.2.3   *Results of Crank–Nicolson scheme.*

3.3   STABILITY

The simple example solved in the previous section illustrated how some time stepping algorithms may prove unstable under certain circumstances. The stability characteristics of the finite difference approximations, as given by equations (3.2.7)–(3.2.10), can be rigorously examined by means of a Fourier analysis. Similarly, the finite element schemes of equations (3.2.11)–(3.2.14) can be rigorously examined by modal decomposition or an eigenvalue analysis. Consider first the stability of the finite difference schemes.

3.3.1 *Fourier stability analysis*

The simplified, one-dimensional, transient heat conduction equation given by
equation (3.2.1) can also be analysed by the method of separation of variables.
In this way, the solution can be expressed as

$$u(x, t) = a_m(t)e^{imx} \tag{3.3.1}$$

where $i^2 = -1$ and $m = $ mode number. Therefore

$$a_m(t) = e^{-m^2 t}a_m(0) \tag{3.3.2}$$

where $a_m(0) = $ initial value of a_m, i.e. at time zero.

Substitution of (3.3.2) into (3.3.1) leads to the solution being expressed as a
Fourier series in the form

$$u(x, t) = \sum_{m=-\infty}^{\infty} a_m(0)e^{-m^2 t}e^{imx} \tag{3.3.3}$$

Therefore, if we require solutions to (3.2.10) in the mth mode of the form

$$u_j^n = (a_m)^n e^{im j\Delta x} \tag{3.3.4}$$

and we let $\mu = \sigma \Delta t/(\Delta x)^2$, then equation (3.2.10) becomes

$$[(a_m)^{n+1} - (a_m)^n]e^{im j\Delta x} = \mu[\theta(a_m)^{n+1}\delta_x^2 e^{im j\Delta x}$$
$$+ (1 - \theta)(a_m)^n \delta_x^2 e^{im j\Delta x}] \tag{3.3.5}$$

But

$$\delta_x^2 e^{im j\Delta x} = e^{im(j+1)\Delta x} - 2e^{im j\Delta x} + e^{im(j-1)\Delta x}$$
$$= e^{im j\Delta x}[e^{im\Delta x} + e^{im\Delta x} - 2]$$
$$= 2e^{im j\Delta x}[\cos m\Delta x - 1]$$
$$= -4\sin^2\left(\frac{m\Delta x}{2}\right)e^{im j\Delta x}$$

Thus

$$(a_m - 1) = -4\mu \sin^2\left(\frac{m\Delta x}{2}\right)[\theta a_m + (1 - \theta)]$$

$$a_m = \frac{1 - 4\mu(1 - \theta)\sin^2\left(\dfrac{m\Delta x}{2}\right)}{1 + 4\mu\theta \sin^2\left(\dfrac{m\Delta x}{2}\right)} \tag{3.3.6}$$

a_m is known as the solution amplification factor. The solution is multiplied by
this factor at the end of each time step. Obviously, for a stable solution we

require that

$$|a_m| < 1$$

Since both σ and μ are greater than (or equal to) zero, and $0 \leqslant \theta \leqslant 1$, then $a_m < 1$. However, we must also ensure that $a_m > -1$. If, for brevity, we let $s = \sin(m\Delta x/2)$, then we must ensure

$$\frac{1 - 4\mu(1 - \theta)s^2}{1 + 4\mu\theta s^2} > -1$$

Noting that the maximum value attainable by s^2 is 1, we then have $2\mu(1 - 2\theta) < 1$.

Obviously, if $1/2 \leqslant \theta \leqslant 1$, then the stability of the solution is ensured. This is therefore the case for both the backward difference scheme, where $\theta = 1$, and the Crank–Nicolson, scheme where $\theta = 1/2$. However, when $0 \leqslant \theta \leqslant 1/2$, then to ensure stability, we must satisfy the condition:

$$\mu = \frac{\Delta t}{\Delta x^2} < \frac{1}{2(1 - 2\theta)}$$

Consequently, any scheme in which $0 < \theta < 1/2$ (such as the forward difference method, $\theta = 0$) is said to be *conditionally stable*. The forward difference scheme is subject to

$$\Delta t < \frac{(\Delta x)^2}{2} \tag{3.3.7}$$

and the previous example demonstrated the instability of the scheme for the larger time steps. Although the Crank–Nicolson solution was not divergent, it did show some oscillatory behaviour for the larger time step. This is due to the fact that when

$$\mu > \frac{1}{A(1 - \theta) \sin^2\left(\dfrac{m\Delta x}{2}\right)} \tag{3.3.8}$$

a_m becomes negative. Therefore, whenever the time step size violates (3.3.8), $(a_m)^n$ changes sign and the solution oscillates. For this reason, the Crank–Nicolson algorithm is often termed as being only *marginally stable* and its oscillations frequently referred to as numerical 'noise'.

The central difference scheme, given by equation (3.2.9), was not used in the example, since it was previously stated that the algorithm was unstable and should never be used. Having carried out the Fourier stability analysis for the alternative schemes, the same method can be applied to the central difference scheme, which will provide a quadratic expression for the amplification factor, given by

$$a_m^2 + (8\mu s^2)a_m - 1 = 0 \tag{3.3.9}$$

It can be seen that the roots of this equation will both be real, one having a modulus greater than unity. Consequently, the solution will always be divergent and the algorithm is said to be *unconditionally stable*. It should never be used, under any circumstances.

3.3.2 Modal decomposition analysis

The most frequently used method of determining the stability of the finite element equations given by (3.2.11)–(3.2.14) is modal decomposition. For the case of free response, $f = 0$, equation (3.1.2) becomes

$$M\dot{T} + kT = 0 \qquad\qquad (3.3.10)$$

The solution to this equation can now be expressed in terms of its linearly independent eigenvectors and eigenvalues by

$$T = \sum_{i=1}^{n} \bar{T}_i e^{\alpha_i t} \qquad\qquad (3.3.11)$$

where $\bar{T}_1 =$ eigenvectors and $\alpha_i =$ eigenvalues.

When $f \neq 0$, it is assumed that the solution can be expressed in terms of the eigenvectors, or modes, as

$$T = \sum_{i=1}^{n} \bar{T}_i y_i(t) \qquad\qquad (3.3.12)$$

where $y_i(t) =$ the mode participation factor.

If this expression is substituted into equation (3.1.2) and then premultiplied by the eigenvectors T_i', for all $i = 1, \ldots, n$, we obtain a set of independent scalar equations given by

$$m_i \dot{y}_i + k_i y_i = f_i \qquad\qquad (3.3.13)$$

where

$$m_i = \bar{T}_i' M T_i$$
$$k_i = \bar{T}_i' K \bar{T}_i$$
$$f_i = \bar{T}_i' f$$

By applying the same method to the general finite element algorithm, equation (3.2.14), for free response $(f = 0)$ we have

$$(m_i + \theta \Delta t k_i)(y_i)_{n+1} = (m_i - (1-\theta)\Delta t k_i)(y_i)_n \qquad (3.3.14)$$

In the previous section the solution at the new time step was expressed in terms of an amplification factor, by which the previous solution was multiplied.

If we now let the amplification factor be given by λ, then

$$(y_i)_{n+1} = \lambda(y_i)_n$$

$$\lambda = \frac{m_i - (1-\theta)\Delta t k_i}{m_i + \theta \Delta t k_i}$$

$$= \frac{1 - (1-\theta)\Delta t(k_i/m_i)}{1 + \theta \Delta t(k_i/m_i)}$$

$$= \frac{1 - (1-\theta)\Delta t \omega_i}{1 + \theta \Delta t \omega_i} \qquad (3.3.15)$$

where $\omega_i = k_i/M_i$ is the eigenvalue of the ith mode of the system. Therefore, as before, in order to ensure stability of the solution, we must satisfy

$$|\lambda| < 1$$

Since $\Delta t \geqslant 0$ and $\omega_i \geqslant 0$ and $0 \leqslant \theta \leqslant 1$, then $\lambda < 1$. Hence, stability can again be ensured by

$$\lambda > -1$$

Therefore

$$(1/2)\omega_i \Delta t(1 - 2\theta) < 1 \qquad (3.3.16)$$

Consequently, for $\theta \geqslant 1/2$, the solution is unconditionally stable, i.e. as for the backward difference and Crank–Nicolson schemes. When $0 \leqslant \theta < 1/2$, the conditional stablility of the solution is provided by

$$\omega_i \Delta t < \frac{2}{1 - 2\theta} \qquad (3.3.17)$$

Hence, the forward difference scheme, with $\theta = 0$, requires a time step limitation to ensure its stability, i.e.

$$\Delta t < \frac{2}{\omega_i} \qquad (3.3.18)$$

Figure 3.3.1 shows a plot of the amplification factor, λ, against $\omega_i \Delta t$. It shows the instability of the forward difference scheme for $\omega_i \Delta t \geqslant 2$, and no oscillations with the backward difference scheme. When $\theta < 1$, λ will become negative for large values of Δt.

Application of modal decomposition to the central difference algorithm, equation (3.2.13), yields a quadratic equation in terms of the amplification factor λ:

$$\lambda^2 + (2\Delta t \omega_i)\lambda - 1 = 0 \qquad (3.3.19)$$

As was the case for the finite difference equations and the Fourier analysis,

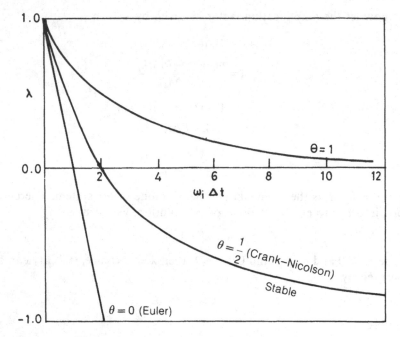

Figure 3.3.1 *Behaviour of amplification factor.*

the roots of this equation are real, with one such that $|\lambda| > 1$. We have once again demonstrated the unconditional instability of the central difference scheme.

3.4 FURTHER DIFFERENCING SCHEMES

The standard finite difference time stepping algorithms have been established and their relative stability characteristics demonstrated in the preceding sections. However, more advanced algorithms have also proved successful in modelling transient heat conduction problems.

Predictor–corrector algorithms [13], for instance, have been adopted for non-linear problems that require an iteration process within each time step. The initial estimate (or prediction) is calculated using a computationally cheap explicit solution, which is subsequently corrected by a more accurate, higher order scheme. The overall computational effort is less and the accuracy of the solution is of the same order as the corrector scheme. Stability of these algorithms is considerably more difficult to determine.

The Lees algorithm [12] provides a three-level scheme which is second-order accurate spatially and temporally and is also unconditionally stable. It is often used in non-linear problems to avoid repeated iterations at each time step. Although unconditionally stable, it may be prone to oscillation.

To reduce the computational expense and effort involved in multi-dimensional problems in inverting matrices with a large bandwidth, the Alternating Direction Method [14] may be utilised. For a two-dimensional problem, the operation is split by first executing a backward difference along the x direction for half a time step, $\Delta t/2$, followed by a similar half-step backward difference in the y direction. Whilst on their own, these schemes are only conditionally stable, their alternate use has been shown to be unconditionally stable [15].

3.5 FINITE ELEMENT TIME STEPPING

We have seen the derivation of finite difference algorithms by approximating the derivatives in time by the finite difference method. In order to produce a finite element algorithm, we shall therefore discretise the temperature in time by the normal finite element procedure, i.e.

$$\mathbf{T} = \sum N_i(t)\mathbf{T}^i \tag{3.5.1}$$

The differential equation given by (3.1.1) is first order, with respect to the time derivative. It is therefore only necessary to provide first-order (i.e. linear) shape functions, $N_i(t)$, in time. It should also be noted at this stage that N_i are assumed to be the same for each component of T and are therefore scalar. If the nodal temperature values change from T^n to T^{n+1} over a time step of length Δt, as shown in Figure 3.5.1, the shape functions are given by

$$N_n = 1 - \xi \qquad N_{n+1} = \xi \tag{3.5.2}$$

where ξ is a local variable varying between 0 and 1 and is given by

$$\xi = \frac{t}{\Delta t}$$

Hence the temporal derivatives of the shape functions are given by

$$\dot{N}_n = \frac{-1}{\Delta t} \qquad \dot{N}_{n+1} = \frac{1}{\Delta t} \tag{3.5.3}$$

The problem can now be developed in two different ways. The weighted residual method can be applied in the normal manner, or the error with respect to T^{n+1} can be minimised to produce a least squares algorithm. We consider each of these methods in turn.

3.5.1 Weighted residual method

The discretisation of the equation system (3.1.2) in time produces

$$\mathbf{M}(\mathbf{T}^n N_n + \mathbf{T}^{n+1} N_{n+1}) + \mathbf{K}(\mathbf{T}^n N_n + \mathbf{T}^{n+1} N_{n+1}) = \mathbf{f} \tag{3.5.4}$$

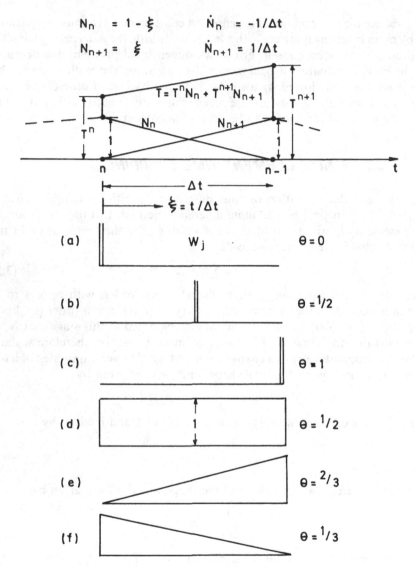

Figure 3.5.1 *Shape and weighting functions for two-level scheme.*

Employing the usual weighted residual method in (3.5.4) then gives

$$\int_0^1 W_j[\mathbf{M}(\mathbf{T}^n N_n + \mathbf{T}^{n+1} N_{n+1}) + \mathbf{K}(\mathbf{T}^n N_n + \mathbf{T}^{n+1} N_{n+1}) - \mathbf{f}]d\xi = 0$$

$$(3.5.5)$$

Substituting in the expressions for the shape functions, (3.5.2), and their derivatives, (3.5.3), then leads to

$$\left(\frac{\mathbf{M}}{\Delta t} + \mathbf{K}\theta\right)\mathbf{T}^{n+1} = \left(\frac{\mathbf{M}}{\Delta t} - \mathbf{K}(1-\theta)\right)\mathbf{T}^n + \hat{\mathbf{f}} \qquad (3.5.6)$$

where

$$\theta = \frac{\int_0^1 W_j \xi \, d\xi}{\int_0^1 W_j \, d\xi} \quad \text{and} \quad \hat{\mathbf{f}} = \frac{\int_0^1 W_j \mathbf{f} \, d\xi}{\int_0^1 W_j \, d\xi}$$

If the same spatial interpolation is used for both $\mathbf{f}$ and $\mathbf{T}$, then $\hat{\mathbf{f}}$ is given by

$$\hat{\mathbf{f}} = \theta \mathbf{f}^n + (1-\theta)\mathbf{f}^{n+1} \qquad (3.5.7)$$

Comparison of the finite element algorithm, (3.5.6), with the previously derived finite difference algorithm, (3.2.14), illustrates that there is no difference between the two schemes when written as matrix systems. Whereas in the finite difference algorithm the choice of θ was completely arbitrary between 0 and 1, its value is now dependent on the weighting function, W_j. Figure 3.5.1(a)–(f) shows how different functions of W_j yield different values for θ. The point collocation method applied at n, $n + 1/2$ and $n + 1$ gives the forward difference scheme ($\theta = 0$) the Crank–Nicolson method ($\theta = 1/2$) and the backward difference method ($\theta = 1$), respectively. The subdomain method with $\theta = 1/2$ is shown in Figure 3.5.1(d). The finite difference schemes are therefore special cases of the weighted residual finite element scheme.

If the shape functions N_n and N_{n+1} are employed for W_j over a single time step, Δt, the so-called Galerkin schemes with $\theta = 1/3$ and $\theta = 2/3$ result. These are shown by Figure 3.5.1(e) and (f). When $\theta = 2/3$, i.e. the weighting employed corresponds to the unknown function, the Galerkin scheme exhibits considerable computational advantage. It is unconditionally stable and gives less oscillatory errors than the Crank–Nicholson scheme, with $\theta = 1/2$.

The stability of the algorithm given by equation (3.5.6) can be examined by using the same procedure detailed for the finite difference schemes in Section 3.3. Three-level time stepping algorithms can also be derived using this method, either by letting $W_j = N_i$ over two time steps, $2\Delta t$, or by employing higher order (e.g. quadratic) shape functions across $2\Delta t$.

3.5.2 Least squares method

A least squares algorithm is derived by utilising a functional that minimises the squares of the error in the solution at the new time step, $\mathbf{T}^{n+1}$. Therefore, across each time step, i.e. between n and $n + 1$, we must minimise the functional

$$\pi = \int_0^1 [\mathbf{M}(\mathbf{T}^n \dot{N}_n + \mathbf{T}^{n+1} \dot{N}_{n+1}) + \mathbf{K}(\mathbf{T}^n N_n + \mathbf{T}^{n+1} N_{n+1}) - \mathbf{f}]^2 \, d\xi \qquad (3.5.8)$$

By noting that, for any of the vectors above, we can write

$$X^2 = X^T \cdot X$$

the resulting least squares scheme is given by

$$\left[\frac{M^T \cdot M}{\Delta t} + \frac{(K^T \cdot M + M^T \cdot K)}{2} + K^T \cdot K \frac{\Delta t}{3} \right] T^{n+1}$$

$$= \left[\frac{M^T \cdot M}{\Delta t} + \frac{(K^T \cdot M - M^T \cdot K)}{2} - K^T \cdot K \frac{\Delta t}{6} \right] T^n$$

$$- K^T \int_0^1 f \xi \frac{d\xi}{\Delta t^2} - M^T \int_0^1 f \frac{d\xi}{\Delta t} \qquad (3.5.9)$$

Although computationally more expensive, due to the greater number of matrix multiplications performed, this algorithm has been demonstrated to exhibit exceptional accuracy [16]. Note, also, that even if the individual matrices, **K** and **M**, are unsymmetric, all the matrix products in the algorithm are symmetric.

3.6 FURTHER FINITE ELEMENT SCHEMES

One of the more complex finite element schemes is the Alternating Direction Galerkin algorithm, which is not entirely analogous to its finite difference namesake [14]. This method takes advantage of the tensor product of certain elements, such as Lagrangian elements, i.e.

$$N_i(x, y) = \phi_i(x), \psi_i(y)$$

The system matrices can then be factorised into directional components, i.e.

$$A = A_x \cdot A_y$$

The matrices that result from this scheme are tridiagonal for linear elements, which produces an efficient solution scheme [17]. In most applications of the method in heat conduction, the basic equations are solved by direct numerical time integration methods. Hence, the computational effort is proportional to the number of unknowns and the number of time steps required to obtain an accurate and stable solution. The size of the problem can be reduced by the introduction of 'thermal modes', which involves the solution of eigenvalue problems. Significant computational effort is involved when solving for the eigenvectors of large systems and selecting a sufficient number of thermal modes that result in an accurate solution is a difficult task. The Lanczos algorithm is used to extract eigenvalues/eigenvectors and this process is very inexpensive [18]. This method will be effective for problems that involve near steady-state solutions or problems in which large number of time steps are required to obtain a stable solution by direct numerical integration. Nour-Amid [19] derived an inexpensive relation to compute proper error bounds to the Lanczos algorithm. Similar approaches have been used for non-linear analysis [20]. Alvaro *et al.* [21], utilising a coordinate transformation matrix generated by the Lanczos algorithm, provide a method for transient heat conduction analysis with costs of the same order as those of steady-state analysis. A finite element eigenvalue method for solving transient heat conduction problems is given by Jia Kang Zhong *et al.* [23], where an analytical solution for the temperature at each finite element node is derived and this is helpful in treating fluid, solid and thermal structure interaction problems. Winget and Hughes [22] give solution algorithms for non-linear transient heat conduction analysis employing element-by-element iterative strategies.

REFERENCES

[1] Zienkiewicz O. C. and Taylor R. L., *The Finite Element Method*, Vol. 1 (4th Edition) McGraw-Hill, New York (1989).

[2] Zienkiewicz O. C. and Morgan K., *Finite Elements and Approximations* John Wiley & Sons, New York (1983).

[3] Myers G. E., *Analytical Methods in Conduction Heat Transfer* McGraw-Hill, New York (1971)

[4] Nagatov E. P., *Applications of Numerical Methods in Heat Transfer* McGraw-Hill, New York (1978).

[5] Bathe K. J., *Finite Element Procedures in Engineering Analysis* Prentice-Hall, Englewood Cliffs, New Jersey (1982).

[6] Bathe K. J. and Wilson E. L., *Numerical Methods in Finite Element Analysis* Prentice-Hall, Englewood Cliffs, New Jersey (1976).

[7] Segerlind L. J., *Applied Finite Element Analysis* (2nd Edition) John Wiley & Sons, New York (1984).

[8] Hughes T. J. R., Unconditionally stable algorithms for nonlinear heat conduction *J. Com. Meth. Appl. Mech. Engg.*, **10**, 135–141 (1977).

[9] Bathe K. J. and Khoshgoftaar M. R., finite element formulation and solution of nonlinear heat transfer *J. Nucl. Engg. Des.*, **51**, 389–401 (1979).

[10] Zienkiewicz O. C. and Lewis R. W., An analysis of various time stepping schemes for initial value problems *Int. J. Earthquake Engg. Struct. Dynamics*, **1**, 407–478 (1975).

[11] Wood W. L. and Lewis R. W., A comparison of time-marching schemes for the transient heat conduction equation *Int. J. Num. Meth. Engg.*, **9**, 679–689 **(1975)**.

[12] Lees M., A linear three-level difference scheme for quasilinear parabolic equations *Math. Comp.*, **20** 516–522 (1966).

[13] Lewis R. W., White I. R. and Wood W. L., A starting algorithm for the numerical simulation of two-phase flow problems, *Int. J. Num. Meth. Engg.*, **12**, 319–329 (1979).

[14] Lewis R. W., Morgan K. and Roberts P. M., Application of an alternating direction finite element method to heat transfer problems involving phase change *Numerical Heat Transfer*, Vol. 7, Hemisphere Press (1984).

[15] Hayes L. J., Implementation of finite element alternating direction methods on non-rectangular regions *Int. J. Num. Meth. Engg.*, **16**, 35–49 (1980).

[16] Lewis R. W. and Bruch J. C. Jr., An application of least squares to one-dimensional transient problems *Int. J. Num. Meth. Engg.*, **8**, 636–647 (1974).

[17] Carnahan B., Luther H. A. and Wilkes J. O., *Applied Numerical Methods* John Wiley & Sons, New York, pp 454–461 (1969).

[18] Nour-Amid B. and Wilson E. L., A new algorithm for heat conduction analysis, *Numerical Methods in Thermal Problems*, eds R. W. Lewis and K. Morgan, Pineridge Press, Swansea, pp. 18–27 (1985).

[19] Nou-Amid B., Lanczos method for heat conduction analysis, *Int. J. Num. Meth. Engg.*, **24**, 251–262 (1987).

[20] Cardona A. and Idelsohn S., Solution of nonlinear thermal transient problems by a reduction method, *Int, J. Num. Meth. Engg.*, **23**, 1023–1042 (1986).

[21] Alvaro, L. G. A., Coutinho, Landau L., Wrobel, Luiz L., and Ebecken, Nelsan F. F., Modal solution of transient heat conduction utilising Lanczos algorithm, *Int. J. Num. Meth. Engg.*, **28**, 13–25 (1989).

[22] Winget J. M. and Hughes T. J. R., Solution algorithms for nonlinear transient heat conduction analysis employing element-by-element iterative strategies *Comp. Math. Appl. Mech., Engg.*, **52**, 711–811 (1985).

[23] Jia Kang Zhong, Chow, Louis C. and Chang, Won Soon, A finite element eigenvalue method for solving transient heat conduction problems, *Int. J. Num. Meth. Heat & Fluid Flow*, **2**, no. 3, 243–260 (1992).

4 Nonlinear Heat Conduction Analysis

4.1 INTRODUCTION

In this chapter we will show how problems involving non-linear transient heat conduction may be solved using simple finite elements and a Galerkin form of the weak formulation of the type introduced in Chapter 1. The method will be extended further by including non-linear boundary conditions.

Transient problems are propagating problems. Knowing the temperature distribution at a particular instant, we are interested to determine the temperature distribution at a later time. Several of the time stepping schemes discussed in the Chapter 3 can be used for this purpose.

4.2 APPLICATION OF GALERKIN'S METHOD TO NON-LINEAR TRANSIENT HEAT CONDUCTION PROBLEMS

4.2.1 Governing equation with initial and boundary conditions

Non-linear transient heat conduction in a stationary medium is governed by

$$\frac{\partial}{\partial x}\left(k(T)\frac{\partial T}{\partial x}\right) + \frac{\partial}{\partial y}\left(k(T)\frac{\partial T}{\partial y}\right) + \frac{\partial}{\partial z}\left(k(T)\frac{\partial T}{\partial z}\right) + Q = \rho c \frac{\partial T}{\partial t} \quad (4.2.1a)$$

where $k(T)$ is a function of the temperature. In terms of the enthalpy, equation (4.2.1a) can be modified to give

$$\frac{\partial}{\partial x}\left[\frac{k(T)}{\rho c}\frac{\partial H}{\partial x}\right] + \frac{\partial}{\partial y}\left[\frac{k(T)}{\rho c}\frac{\partial H}{\partial y}\right] + \frac{\partial}{\partial z}\left[\frac{k(T)}{\rho c}\frac{\partial H}{\partial z}\right] + Q = \rho c \frac{\partial H}{\partial t} \quad (4.2.1b)$$

where

$$H_2 - H_1 = \int_{T_1}^{T_2} \rho c \, dT \quad (4.2.1c)$$

The boundary conditions of the problem are

$$T = T_b \quad \text{on } \Gamma_b \tag{4.2.2a}$$

$$k(T)\frac{\partial T}{\partial x}l_x + k(T)\frac{\partial T}{\partial y}l_y + k(T)\frac{\partial T}{\partial z}l_z + q + \alpha(T - T_\infty) = 0 \quad \text{on } \Gamma_q \tag{4.2.2b}$$

where l_x, l_y and l_z are direction cosines of outward normal, $\alpha =$ heat transfer coefficient, $T_\infty =$ ambient temperature. The initial condition for the problem is

$$T = T_0 \quad \text{at } t = 0 \tag{4.2.2c}$$

4.2.2 Galerkin's method

Galerkin's approach is adopted here to solve equation (4.2.1), subject to the various conditions of equations (4.2.2a)–(4.2.2c). The solution domain is divided into finite elements in space. The temperature is approximated within each element by

$$T(x, y, z, t) = \sum_{i=1}^{m} N_i(x, y, z)T(t) \tag{4.2.3}$$

where N_i are the usual shape functions defined piece-wise, element by element, $T(t)$ is the nodal temperatures considered to be function of time and m is the number of nodes in the element considered.

The Galerkin representation for the heat conduction problem (equation (4.2.1)) is

$$\int N_i \left[\frac{\partial}{\partial x}\left(k_x(T)\frac{\partial T}{\partial x}\right) + \frac{\partial}{\partial y}\left(k_y(T)\frac{\partial T}{\partial y}\right) + \frac{\partial}{\partial z}\left(k_z(T)\frac{\partial T}{\partial z}\right) + Q - \rho c\frac{\partial T}{\partial t} \right] dx\, dy\, dz = 0 \tag{4.2.4}$$

Using integration by parts on the first three terms in equation (4.2.4), the equation simplifies to

$$-\int \left[k_x(T)\frac{\partial T}{\partial x}\frac{\partial N_i}{\partial x} + k_y(T)\frac{\partial T}{\partial y}\frac{\partial N_i}{\partial y} + k_z(T)\frac{\partial T}{\partial z}\frac{\partial N_i}{\partial z} - N_i Q - N_i \rho c\frac{\partial T}{\partial t} \right] dx\, dy\, dz$$

$$-\int N_i q\, d\Gamma_q - \int N_i \alpha(T - T_\infty)\, d\Gamma_q = 0 \quad i = 1, 2, \ldots, m \tag{4.2.5}$$

Inserting the temperature approximation, equation (4.2.5) simplifies to

$$-\int \left[k_x(T)\frac{\partial N_j}{\partial x}\frac{\partial N_i}{\partial x}\{T\} + k_y(T)\frac{\partial N_j}{\partial y}\frac{\partial N_i}{\partial y}\{T\} + k_z(T)\frac{\partial N_j}{\partial z}\frac{\partial N_i}{\partial z}\{T\} \right] dx\, dy\, dz$$

$$-\int N_i \alpha \cdot N_j \{T\}\, dT_\alpha + \int N_i Q\, dx\, dy\, dz - \int N_i \rho c N_j\, dx\, dy\, dz \frac{\partial\{T\}}{\partial t} \tag{4.2.6}$$

$$-\int N_i q\, d\Gamma_q + \int N_i 2T_\infty\, d\Gamma_\alpha = 0$$

The above equation, (4.2.6), can be cast into a more convenient form as

$$\mathbf{M}\frac{d\mathbf{T}}{dt} + \mathbf{KT} = \mathbf{f} \tag{4.2.7}$$

where

$$M_{ij} = \int \rho c N_i N_j \, dx \, dy \, dz$$

$$K_{ij} = \int \left[k_x(T) \frac{\partial N_i}{\partial x} \frac{\partial N_j}{\partial x} + k_y(T) \frac{\partial N_i}{\partial y} \frac{\partial N_j}{\partial y} \right.$$
$$\left. + k_z(T) \frac{\partial N_i}{\partial z} \frac{\partial N_j}{\partial z} \right] dx \, dy \, dz + \int \alpha N_i N_j \, d\Gamma_\alpha$$

$$f_i = \int N_i Q \, dx \, dy \, dz - \int q N_i \, d\Gamma_q + \int N_i \alpha T_\infty \, d\Gamma_\alpha$$

This non-linear equation set requires an iterative solution. Following the simplest form of iteration method we could start from some initial guess:

$$T = T^0 = (T_1^0, T_2^0, \cdots T_m^0) \tag{4.2.8}$$

and obtain an improved solution T^1 by solving the equation

$$\mathbf{M} \, d\mathbf{T}^1/dt + \mathbf{K}(\mathbf{T}^0)\mathbf{T}^1 = \mathbf{f}^0 \tag{4.2.9}$$

The general iteration scheme

$$\mathbf{M}d\mathbf{T}^n/dt + \mathbf{K}(\mathbf{T}^{n-1})\mathbf{T}^n = \mathbf{f}^{n-1} \tag{4.2.10}$$

is then repeated until convergence, to within a suitable tolerance, is obtained.

4.2.3   One-dimensional non-linear steady-state problem

Example 4.1.
The above relations are illustrated, first, for the case of a non-linear steady-state problem using a linear one-dimensional element. Example 2.1 is reconsidered here except that the thermal conductivity is replaced by $k = (10 + 0.1T)$. One face is maintained at a temperature of $T = 5$, while heat flux of $q = 3T - 7$ is maintained at the other face.

The distribution of the temperature satisfies the equation

$$(10 + 0.1T)(d^2 T/dx^2) + 180x = 0 \quad 0 < x < 1$$

and the boundary conditions

$$T = 5 \qquad \text{at } x = 0$$
$$q = 3T - 7 \quad \text{at } x = 1$$

Using a single element to represent the region, with node 1 at $x = 0$ and node 2 at $x = 1$, the element equations become

$$(10 + 0.1T)\begin{bmatrix} 1 & -1 \\ -1 & 1 \end{bmatrix}\begin{Bmatrix} T_1 \\ T_2 \end{Bmatrix} = \frac{1}{6}\begin{bmatrix} 2 & 1 \\ 1 & 2 \end{bmatrix}\begin{Bmatrix} Q_1 \\ Q_2 \end{Bmatrix} - \begin{Bmatrix} q_1 \\ q_2 \end{Bmatrix}$$

where, by definition, $Q_1 = 0$ and $Q_2 = 180$. The equation corresponding to node 2 is then

$$(10 + 0.1T)(-T_1 + T_2) = \frac{2(180)}{6} - q_2$$

$T_1 = 5$ and $q_2 = 3T_2 - 7$ gives

$$(10 + 0.1T)(-5 + T_2) = 60 - 3T_2 + 7 \tag{4.2.10a}$$

In order to solve this equation, we should know the value of T. Since a linear element is used, T can be represented by $(T_1 + T_2)/2$. The equation to be solved is given by

$$\left[10 + 0.1\left(\frac{5 + T_2}{2}\right)\right](-5 + T_2) = 67 - 3T_2$$

As a first approximation, let us consider the value of $T_2 = 9$ given earlier for the linear problem where $k = 10$ units:

$$\text{LHS}\left[10 + 0.1\left(\frac{5 + 9}{2}\right)\right][-5 + 9] = (10 + 0.07)4$$

$$= 40.28$$

$$\text{RHS} = 6T - 3(9) = 67 - 27 = 40 \text{ units}$$

Therefore the assumed value is not satisfactory. Simplifying equation (4.2.10a) as $K(T_0)(-5 + T_2^1) = f^0$, where

$$K(T_0) = \left[10 + 0.1\left(\frac{T_1 + T_2}{2}\right)\right]$$

$$f^0 = (67 - 3T_2^0)$$

we can form a recurrence relationship and determine T_2 to the accuracy required. Let $T_2 = 9.1$,

$$K(T_0) = 10 + 0.1\left(\frac{5 + 9.1}{2}\right) = (10 + 0.705)$$

$$f_0 = 67 - 27.3 = 39.7$$

$$-5 + T_2^0 = \frac{39.7}{10.705} = 3.63$$

Therefore

$$T_2^1 = 8.63$$

$$K(T_1) = 10 + 0.1 \left(\frac{5 + 8.63}{2} \right) = 10.681$$

$$f_1 = 67 - 3(8.631) = 672649 = 40.51$$

$$5 + T_2^2 = \frac{40.51}{10.681} = 3.79$$

Therefore

$$T_2 = 5 + 3.79 = 8.79$$

$$K(T_2) = 10 + 0.1 \left(\frac{5 + 8.79}{2} \right) = 10.69$$

$$f_2 = 67 - 3(8.79) = 67 - 26.37 = 40.73$$

$$T_2 = 5 + 3.81 = 8.81$$

Thus the solution can be improved with more iterations.

4.2.4   One-dimensional transient state problem

The relation derived in equation (4.2.7) will now be considered to illustrate the application to a one-dimensional transient problem using a linear element. T is represented in the element by

$$T = N_i T_i + N_j T_j \tag{4.2.11}$$

where

$$N_i = 1 - x/L \quad \text{and} \quad N_j = x/L \tag{4.2.12}$$

$$\frac{dT}{dx} = \frac{dN_i}{dx} T_i + \frac{dN_j}{dx} T_j$$

$$= -(1/L)T_i + (1/L)T_j$$

$$[M] = \int_L \rho c A \begin{bmatrix} N_i^2 & N_i N_j \\ N_i N_j & N_j^2 \end{bmatrix} dx = \frac{\rho c L A}{6} \begin{bmatrix} 2 & 1 \\ 1 & 2 \end{bmatrix} \tag{4.2.13}$$

$$[K] = \int_L K_x(T) \frac{dN_i}{dx} \frac{dN_j}{dx} A \, dx + \int_{\Gamma_a} \alpha N_i N_j d\Gamma_a$$

$$= \int_L \frac{A K_x(T)}{L} \begin{bmatrix} \left(\frac{1}{L^2}\right) & -\left(\frac{1}{L^2}\right) \\ -\left(\frac{1}{L^2}\right) & \left(\frac{1}{L^2}\right) \end{bmatrix} dx = A K_x/L \begin{bmatrix} 1 & -1 \\ -1 & 1 \end{bmatrix} + \frac{\alpha L P}{6} \begin{bmatrix} 2 & 1 \\ 1 & 2 \end{bmatrix}$$

$$\tag{4.2.14}$$

$$f_i = \int_L N_i QA\,dx - \int_{\Gamma_q} qN_i\,d\Gamma_q + \int_{\Gamma_x} \alpha T_\infty N_i\,d\Gamma_\alpha$$

$$= \frac{AQL}{2}\begin{Bmatrix}1\\1\end{Bmatrix} - qA\begin{Bmatrix}0\\1\end{Bmatrix} + \frac{\alpha T_\infty LP}{2}\begin{Bmatrix}1\\1\end{Bmatrix} \qquad (4.2.15)$$

where Q is constant, P is a perimeter and q is constant. If Q is a linearly varying function, then Q is expressed by $Q = N_i Q_i + N_j Q_j$. Then

$$\int_L N_i[N_i Q_j]\begin{Bmatrix}Q_i\\Q_j\end{Bmatrix} A\,dx = AL/6 \begin{bmatrix}2 & 1\\1 & 2\end{bmatrix}\begin{Bmatrix}Q_1\\Q_2\end{Bmatrix} \qquad (4.2.16)$$

Example 4.2
Consider Example 4.1, where the temperature at the left end of the rod is raised to 10 units instead of 5 units. We are interested to find out the temperature distribution in the rod as a function of space and time.

The initial temperature distribution, at time $t = 0$, is given by $T_{x=0} = 5.0$ and $T_{x=1} = 8.81$. Let $\rho c = 6.0$, $A = 1.0$, $K_x = (10 + 0.1T)$ and $q = 3T - 7$ at node 2.

For the purpose of illustration, let us take one linear element and use the θ-method for the time stepping as mentioned in Chapter 3:

$$(M + \theta\Delta K)T^{n+1} = (M - (1-\theta)\Delta t K)T^n + \Delta t f^n$$

Therefore

$$M = \frac{\rho c A L}{6}\begin{bmatrix}2 & 1\\1 & 2\end{bmatrix} = \begin{bmatrix}2 & 1\\1 & 2\end{bmatrix}$$

$$K = \frac{AK_x(T)}{L}\begin{bmatrix}1 & -1\\-1 & 1\end{bmatrix} + \alpha A\begin{bmatrix}0 & 0\\0 & 1\end{bmatrix}$$

In order to calculate the thermal conductivity, we need to know the temperature distribution. Conversely, in order to determine the temperature distribution, we should know the thermal conductivity. Thus an iterative solution is required to determine the temperature distribution. To start with, we shall use the temperature distribution at the steady state, i.e. $t = 0$, to determine the thermal conductivity. Thus

$$K_x(T) = 10 + 0.1\left(\frac{5 + 8.81}{2}\right) = 10.69$$

$$K = 10.69\begin{bmatrix}1 & -1\\-1 & 1\end{bmatrix} + 3\begin{bmatrix}0 & 0\\0 & 1\end{bmatrix} = \begin{bmatrix}10.69 & -10.69\\-10.69 & 13.69\end{bmatrix}$$

$$f = \frac{AL}{6}\begin{bmatrix}2 & 1\\1 & 2\end{bmatrix}\begin{Bmatrix}Q_1\\Q\end{Bmatrix} + \alpha T_\infty A\begin{Bmatrix}0\\1\end{Bmatrix}$$

$$= \frac{1}{6}\begin{bmatrix}2 & 1\\1 & 2\end{bmatrix}\begin{Bmatrix}0\\180\end{Bmatrix} + 3\left(\frac{7}{3}\right)\begin{Bmatrix}0\\1\end{Bmatrix} = \begin{Bmatrix}30\\67\end{Bmatrix}$$

Let us assume a time step of 0.01 and $\theta = 1/2$, corresponding to the Crank–Nicolson method. Then

$$(\mathbf{M} + \tfrac{1}{2}0.01\mathbf{K})\mathbf{T}^{n+1} = (\mathbf{M} - \tfrac{1}{2}0.01\mathbf{K})\mathbf{T}^n + 0.01\mathbf{f}^n$$

$$\mathbf{M} + \tfrac{1}{2}0.01\mathbf{K} = \begin{bmatrix} 2 & 1 \\ 1 & 2 \end{bmatrix} + \tfrac{1}{2}0.01 \begin{bmatrix} 10.69 & -10.69 \\ -10.69 & 13.69 \end{bmatrix}$$

$$= \begin{bmatrix} 2.054\,35 & 0.946\,55 \\ 0.946\,55 & 2.068\,45 \end{bmatrix} = \mathbf{A}$$

$$\mathbf{M} - \tfrac{1}{2}0.01\mathbf{K} = \begin{bmatrix} 2 & 1 \\ 1 & 2 \end{bmatrix} - \tfrac{1}{2}0.01 \begin{bmatrix} 10.69 & -10.69 \\ -10.69 & 13.69 \end{bmatrix}$$

$$= \begin{bmatrix} 1.946\,55 & 1.053\,45 \\ 1.053\,45 & 1.931\,55 \end{bmatrix} = \mathbf{B}$$

The system of equations to be solved is

$$\begin{bmatrix} 2.053\,45 & 0.946\,55 \\ 0.946\,55 & 2.068\,45 \end{bmatrix} \begin{Bmatrix} T_1 \\ T_2 \end{Bmatrix}^{n+1} = \begin{bmatrix} 1.946\,55 & 1.053\,45 \\ 1.053\,45 & 1.931\,55 \end{bmatrix} \begin{Bmatrix} T_1 \\ T_2 \end{Bmatrix}^n + 0.01 \begin{Bmatrix} 30 \\ 67 \end{Bmatrix}$$

At time $t = 0$, the initial temperatures at nodes 1 and 2 are, respectively, given by 5 and 8.81.

$$\mathbf{A} \begin{Bmatrix} T_1 \\ T_2 \end{Bmatrix} = \begin{bmatrix} 1.946\,55 & 1.053\,45 \\ 1.053\,45 & 1.931\,55 \end{bmatrix} \begin{Bmatrix} 5 \\ 8.81 \end{Bmatrix} + 0.01 \begin{Bmatrix} 30 \\ 67 \end{Bmatrix} = \begin{Bmatrix} 19.313\,64 \\ 22.953\,20 \end{Bmatrix}$$

But the temperature at node 1 should be maintained at 10 for $t > 0$. Hence

$$\begin{bmatrix} 1 & 0 \\ 0 & 2.068\,45 \end{bmatrix} \begin{Bmatrix} T_1 \\ T_2 \end{Bmatrix}_{0.01} = \begin{Bmatrix} 10 \\ 22.953\,20 - 10(0.946\,55) \end{Bmatrix}$$

Therefore

$$T_2 = \frac{22.953\,20 - 10(0.946\,55)}{2.068\,45} = 6.521$$

The temperature T_2 at node 2 is calculated based on the thermal conductivity value determined with an assumed value of 5 and 8.81 at nodes 1 and 2, respectively. Since the problem is non-linear by virtue of the thermal conductivity being a function of temperature, it is necessary to recalculate the temperature at the nodes with the new set of temperatures being used to determine the value of the thermal conductivity

$$K_x(T) = 10 + 0.1 \left(\frac{10 + 6.521}{2} \right) = 10.826$$

Thus **A** and **B** will have to be recalculated:

$$\mathbf{M} + \tfrac{1}{2}0.01\mathbf{K} = \begin{bmatrix} 2 & 1 \\ 1 & 2 \end{bmatrix} + \tfrac{1}{2}0.01 \begin{bmatrix} 10.826 & -10.826 \\ -10.826 & 13.826 \end{bmatrix}$$

$$= \begin{bmatrix} 2.054\,13 & 0.945\,87 \\ 0.945\,87 & 2.069\,13 \end{bmatrix} = \mathbf{A}_1$$

$$\mathbf{M} + \tfrac{1}{2}0.01\mathbf{K} = \begin{bmatrix} 2 & 1 \\ 1 & 2 \end{bmatrix} + \tfrac{1}{2}0.01 \begin{bmatrix} 10.826 & -10.826 \\ -10.826 & 13.826 \end{bmatrix}$$

$$= \begin{bmatrix} 1.945\,88 & 1.054\,12 \\ 1.054\,12 & 1.930\,88 \end{bmatrix} = \mathbf{B}_1$$

The system of equations to be solved is

$$\begin{bmatrix} 2.054\,13 & 0.945\,87 \\ 0.945\,87 & 2.069\,13 \end{bmatrix} \begin{Bmatrix} T_1 \\ T_2 \end{Bmatrix}_{0.01} = \begin{bmatrix} 1.945\,88 & 1.054\,12 \\ 1.054\,12 & 1.930\,88 \end{bmatrix} \begin{Bmatrix} T_1 \\ T_2 \end{Bmatrix}_0 + 0.01 \begin{Bmatrix} 30 \\ 67 \end{Bmatrix}$$

At time $t = 0$, the initial temperatures are given by 5 and 8.81 at nodes 1 and 2, respectively.

$$\mathbf{A}_1 \begin{Bmatrix} T_1 \\ T_2 \end{Bmatrix}_{0.01} = \begin{bmatrix} 1.945\,88 & 1.054\,12 \\ 1.054\,12 & 1.930\,88 \end{bmatrix} \begin{Bmatrix} 5 \\ 8.81 \end{Bmatrix} + 0.01 \begin{Bmatrix} 30 \\ 67 \end{Bmatrix} = \begin{Bmatrix} 19.3162 \\ 22.958\,17 \end{Bmatrix}$$

i.e.

$$\begin{bmatrix} 2.054\,13 & 0.948\,57 \\ 0.948\,57 & 2.069\,13 \end{bmatrix} \begin{Bmatrix} T_1 \\ T_2 \end{Bmatrix}_{0.01} = \begin{Bmatrix} 19.3162 \\ 22.958\,17 \end{Bmatrix}$$

But the temperature at node 1 is to be maintained at 10 for $t > 0$. Hence

$$\begin{bmatrix} 1 & 0 \\ 0 & 2.069\,13 \end{bmatrix} \begin{Bmatrix} T_1 \\ T_2 \end{Bmatrix}_{0.01} = \begin{Bmatrix} 10 \\ 22.958\,17 - 10(0.945\,87) \end{Bmatrix}$$

Therefore T_2 at 0.01 s is given by

$$T_2 = \frac{22.958\,17 - 10(0.945\,89)}{2.069\,13} = 6.524$$

Since the problem is non-linear, the solution is obtained in each time step by iteration. After obtaining the converged solution at the end of the time interval 0.01 s (in this case 6.524), the process is continued to obtain the temperature values at the end of 0.02 s using the values of temperatures at nodes 1 and 2 at 0.01 s.

$$K_x(T) = 10.0 + 0.1 \left(\frac{10 + 6.524}{2} \right) = 10.8962$$

Since $K_x(T)$ does not vary much, we can make use of the values of $\mathbf{A}_1$ and $\mathbf{B}_1$ calculated earlier. The system of equations to be solved to get the temperature

T_2 at the end of 0.02 s is

$$\begin{bmatrix} 2.054\,13 & 0.945\,87 \\ 0.945\,87 & 2.069\,13 \end{bmatrix} \begin{Bmatrix} T_1 \\ T_2 \end{Bmatrix}_{0.02} = \begin{bmatrix} 1.945\,88 & 1.054\,12 \\ 1.054\,12 & 1.930\,88 \end{bmatrix} \begin{Bmatrix} T_1 \\ T_2 \end{Bmatrix}_{0.01} + 0.01 \begin{Bmatrix} 30 \\ 87 \end{Bmatrix}$$

$$= \begin{bmatrix} 1.945\,88 & 1.054\,12 \\ 1.054\,12 & 1.930\,88 \end{bmatrix} \begin{Bmatrix} 10 \\ 6.524 \end{Bmatrix} + 0.01 \begin{Bmatrix} 30 \\ 67 \end{Bmatrix}$$

$$= \begin{Bmatrix} 26.6358 \\ 23.1483 \end{Bmatrix}$$

But T_1 should be maintained at 10 for $t > 0$

$$\begin{bmatrix} 1 & 0 \\ 0 & 2.069\,13 \end{bmatrix} \begin{Bmatrix} T_1 \\ T_2 \end{Bmatrix}_{0.02} = \begin{Bmatrix} 10 \\ 23.1483 - 10(0.945\,87) \end{Bmatrix}$$

Hence T_2 at the end of 0.02 s is given by

$$T_2 = \frac{23.1483 - 10(0.945\,87)}{2.069\,13} = 6.61$$

It can be observed that the value of T_2 rises gradually as time elapses.

With this value of T_2, i.e. 6.61, the thermal conductivity is to be calculated and a new value of T_2 is to be determined. The iteration is to be continued until the converged value of T_2 is obtained within the accuracy criterion used.

The marching process in time is continued until we obtain the temperatures at the desired time. Thus, a non-linear transient problem takes considerably more time and effort for its solution compared with either linear steady or transient problem.

4.3 EXAMPLES

Several examples are considered here in order to illustrate the application of the finite element method to the transient problems for both linear and non-linear cases. The treatment of various boundary conditions that occur in practice is also considered through examples. One-, two-, and three-dimensional examples are considered here for illustration of the material presented in earlier chapters.

To illustrate the solution technique on a transient heat conduction problem for which the exact solution is known, the one-dimensional example of constant heat flux applied to a semi-infinite solid is chosen [1]. The surface of the solid is chosen to be the origin. The material properties and the applied heat flux per unit time are all considered to be unity. The exact solution is given by

$$T(x, t) = 2(t/\pi)^{1/2} \left[\exp(-x^2/4t) - (1/2)x \sqrt{\frac{\pi}{t}} \, \mathrm{erfc}\left(\frac{x}{2\sqrt{t}}\right) \right] \qquad (4.3.1)$$

If the temperatures and the nodal loads vary linearly with time during an arbitrary time interval Δt, the integration equation

$$C\dot{T} + KT = f(t)$$

between time $t - \Delta t$ and t yields the recursion formula (Crank–Nicolson)

$$\left(\tfrac{1}{2}K + \frac{1}{\Delta t}C\right)T(t) = \left(-\tfrac{1}{2}K + \frac{C}{\Delta t}\right)T(t - \Delta t)$$

$$+ \tfrac{1}{2}(F(t - \Delta t) + F(t)) \qquad (4.3.2)$$

This relation coincides with the recurrence formula derived by Wilson and Nickell [2] on the basis of a variational principle due to Gurtin. A similar relation has been used by Zienkiewicz and Parekh [3]. The spatial temperature distributions for various times are plotted in Figure 4.3.1. Except at early times the agreement with the exact solution is excellent. A coarse mesh solution at time $t = 1.0$ is given in Figure 4.3.2. Also, Figure 4.3.2 shows that the effect of diagonalising the heat capacity matrix reduces by almost 50% the number of numerical operations in the solution; however, the loss of accuracy appears to be small due to this approximation [2]. As an alternative formulation, the Galerkin process can be applied to the matrix differential equation and various integration schemes derived. Assuming a linear variation of temperatures and nodal loads between $t - \Delta t$ and t, this process yields the following recurrence formula [4]:

$$\left(\tfrac{1}{3}K + \frac{1}{2\Delta t}C\right)T(t) = \left(-\tfrac{1}{6}K + \frac{C}{2\Delta t}\right)T(t - \Delta t)$$

$$+ \tfrac{1}{6}(F(t - \Delta t) + 2F(t)) \qquad (4.3.3)$$

Figure 4.3.3 compares both numerical solutions with the exact one. As can be seen, a much better short-time accuracy is achieved with the finite element solution based on the Galerkin process. Due to stability, the rather severe oscillations that are present in Crank–Nicolson type solutions are damped out for long times. Table 4.3.1 shows the solutions to the example considered for various time intervals. A better short-time accuracy of the Galerkin scheme over the Crank–Nicolson scheme is also observed for two- and three-dimensional problems [4]. Donea [4] also discusses the application of quadratic interpolation over the successive time steps along with the Galerkin method and gives the following recurrence relations:

$$\left(\frac{1}{15}\begin{bmatrix} 8K & K \\ K & 2K \end{bmatrix} + \frac{1}{12\Delta t}\begin{bmatrix} C & 4C \\ -4C & 8C \end{bmatrix}\right)\begin{Bmatrix} T(t) \\ T(t + \Delta t) \end{Bmatrix}$$

$$= \left(-\frac{1}{15}\begin{bmatrix} K \\ -1/2K \end{bmatrix} + \frac{1}{12\Delta t}\begin{bmatrix} 4C \\ -C \end{bmatrix}\right)\{T(t - \Delta t)\}$$

$$+ \frac{1}{15}\begin{bmatrix} 1 & 8 & 1 \\ -1/2 & 1 & 2 \end{bmatrix}\begin{Bmatrix} F(t - \Delta t) \\ f(t) \\ F(t + \Delta t) \end{Bmatrix} \qquad (4.3.4)$$

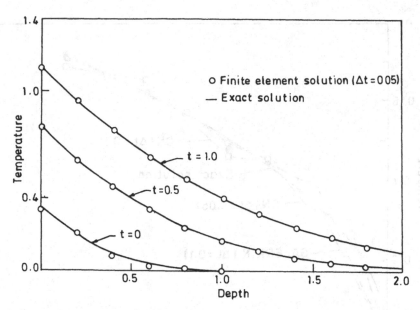

Figure 4.3.1 *Temperature distribution at various times versus depth. (Reproduced by permission from Wilson E. L. and Nickell R. E., Application of the finite element method to heat conduction analysis. Nucl. Engg. Des., 4, 276–286 (1966), © Elsevier Sequoia Lausanne.)*

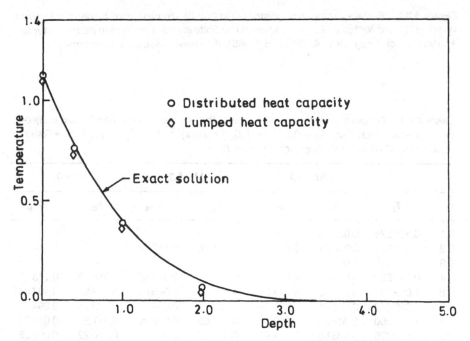

Figure 4.3.2 *Temperature distribution at time t = 1.0 indicating the effect of lumped and distributed heat capacity (with four elements). (Reproduced by permission from Wilson E. L. and Nickell R. E., Application of the finite element method to heat conduction analysis. Nucl. Engg. Des., 4, 276–286 (1966), © Elsevier Sequoia Lausanne.)*

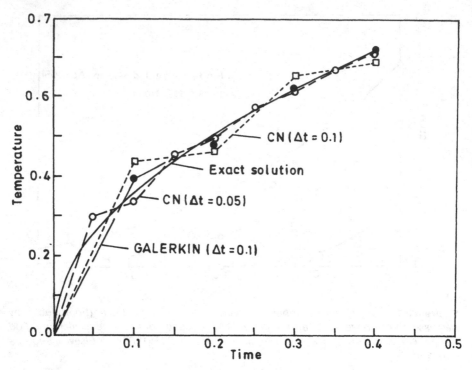

Figure 4.3.3 *Surface temperature distribution (x = 0). (Reproduced by permission from Wilson E. L. and Nickell R. E., Application of the finite element method to heat conduction analysis. Nucl. Engg. Des., 4, 276–286 (1966), © Elsevier Sequoia Lausanne.)*

Table 4.3.1 Temperature variation with time at the origin of a semi-infinite solid subjected to a constant heat flux. $\Delta x = 0.2$; $T_E = T(0, t) = 2(t/\pi)^{1/2}$; $E = |T_E - T_{Num}|$; G = Galerkin scheme (4); C–N = Crank–Nicolson scheme (2)

t	T_E	\multicolumn{2}{c}{$\Delta t = 0.1$}		\multicolumn{2}{c}{$\Delta t = 0.2$}	

t	T_E	E_G	E_{C-N}	E_G	E_{C-N}	E_G	E_{C-N}
0.1	0.356 825	0.021 14	0.076 19	—	—	—	—
0.2	0.504 627	0.018 67	0.044 55	0.036 38	0.117 54	—	—
0.3	0.618 039	0.004 81	0.034 02	—	—	—	—
0.4	0.713 650	0.008 96	0.025 81	0.027 50	0.071 40	0.056 15	0.173 50
0.8	1.009 253	0.005 89	0.011 31	0.012 32	0.046 81	0.039 87	0.107 64
1.2	1.236 078	0.004 91	0.005 75	0.008 90	0.034 08	0.003 76	0.094 88
1.6	1.427 300	0.004 30	0.003 19	0.007 62	0.026 09	0.017 31	0.075 24
2.0	1.595 770	0.003 88	0.001 92	0.006 87	0.020 58	0.010 22	0.068 35
2.4	1.748 078	0.003 56	0.001 25	0.006 34	0.016 55	0.012 16	0.058 42
2.8	1.888 140	0.003 30	0.000 89	0.005 91	0.013 49	0.010 31	0.053 81

Columns under headers: $\Delta t = 0.1$ (E_G, E_{C-N}); $\Delta t = 0.2$ (E_G, E_{C-N}); $\Delta t = 0.4$ (E_G, E_{C-N})

Starting from a known situation at $(t - \Delta t)$, these simultaneous equations furnish the temperature distribution at times t and $t + \Delta t$. Donea [4] gives the numerical values of the temperature at the surface for space intervals $\Delta x = 0.2$ and various time steps Δt for the previously considered case of the one-dimensional example with a constant surface heat flux to a semi-infinite solid.

As an illustration of a two-dimensional problem we consider a transient heat flow in a prismatic fuel element for a nuclear reactor. The general configuration of the fuel element can be seen from Figure 4.3.4 together with the symmetric portion of its cross-section. Figure 4.3.5 shows the finite element grid that has been used in solving this two-dimensional problem. A constant heat flux is assumed to be applied at time $t = 0$ along the interface A–C between the fuel and canning. The other boundary conditions as well as the physical constants of the material are indicated in Figure 4.3.5.

An accurate reference solution to this problem was first derived by using very small time intervals ($\Delta t = 0.01$). Then, the problem was successively solved with $\Delta t = 0.1$, 0.2 and 0.4 by means of both the Galerkin and the Crank–Nicolson schemes. Table 4.3.2 compares the results obtained with these two integration methods to the reference numerical solution. Here again the Galerkin scheme is found to yield a better short-time accuracy than the Crank–Nicolson scheme.

It is expected that higher order time elements would permit an increase in the element width Δt. To justify a change in the boundary conditions or the heat input within the time integration range, these variables were also shown as an interpolation function. This is of special significance in periodic processes [5].

As will be noted from Figure 4.3.6, many periodic processes can be covered

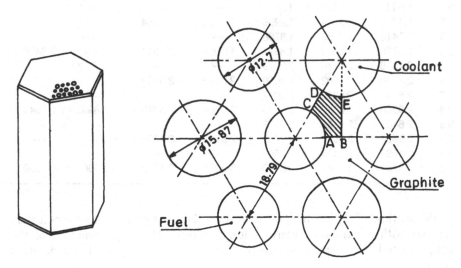

Figure 4.3.4 *Prismatic fuel element for a nuclear reactor and symmetric portion of its cross-section.*

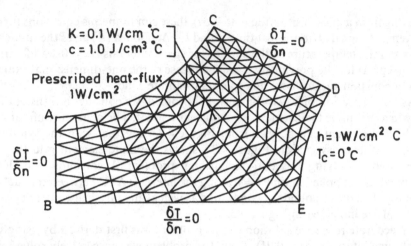

Figure 4.3.5 *Discretisation and boundary conditions for the transient heat flow of the fuel element.*

Table 4.3.2 Temperature variation with time at point C for a constant heat flux applied at $t = 0$ along the interface A–C between fuel and canning. T_R = reference numerical solution ($\Delta t = 0.01$); G = Galerkin scheme (4); C–N = Crank–Nicolson scheme (2)

		$\Delta t = 0.1$		$\Delta t = 0.2$		$\Delta t = 0.4$	
t	T_R	T_G	T_{C-N}	T_G	T_{C-N}	T_G	T_{C-N}
0.1	1.0447	1.1328	1.3114	—	—	—	—
0.2	1.4471	1.3944	1.2996	1.5783	1.8364	—	—
0.3	1.7418	1.7383	1.8758	—	—	—	—
0.4	1.9820	1.9613	1.8877	1.9028	1.7514	2.1774	2.5426
0.5	2.1888	2.1776	2.2779	—	—	—	—
0.6	2.3730	2.3602	2.3059	2.3743	2.5838	—	—
0.7	2.5414	2.5312	2.6057	—	—	—	—
0.8	2.6980	2.6882	2.6476	2.6708	2.5379	2.5893	2.3541
0.9	2.8455	2.8366	2.8937	—	—	—	—
1.0	2.9856	2.9972	2.9470	2.9732	3.1339	—	—
1.1	3.1197	3.1115	3.1568	—	—	—	—
1.2	3.2484	3.2404	3.2186	3.2305	3.1243	3.2530	3.5559

by this method. The time integration is demonstrated by the example of a time element with a parabolic interpolation function. The approximation of the temperature in the time element is (Figure 4.3.7 shows a possible temperature curve)

$$\mathbf{T} = N_0\mathbf{T}_0 + N_1\mathbf{T}_1 + N_2\mathbf{T}_2 \qquad (4.3.5)$$

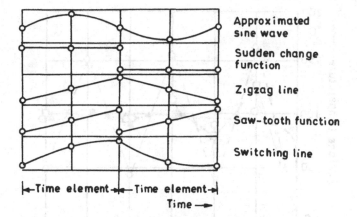

Approximated
sine wave

Sudden change
function

Zigzag line

Saw-tooth function

Switching line

|←Time element→|←Time element→|

Time →

Figure 4.3.6 *Examples of reproduction of periodic changes of boundary conditions.*

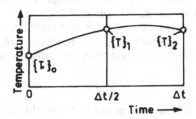

Figure 4.3.7 *Approximation of temperature in time element.*

where

$$N_0 = 1 - 3\frac{t}{\Delta t} + 2\left(\frac{t}{\Delta t}\right)^2$$

$$N_1 = 4\frac{t}{\Delta t} - 4\left(\frac{t}{\Delta t}\right)^2 \qquad N_2 = \frac{-t}{\Delta t} + 2\left(\frac{t}{\Delta t}\right)^2 \qquad (4.3.6)$$

To demonstrate the convergence of the different time elements, extremely large intervals are selected. The surface temperature of a plane plate is reproduced, the temperature of which until time $t = 0$ is constant at $0\,°C$. From that time on an ambient temperature of $100\,°C$ acts on both plate surfaces. The heat transfer coefficient is constant both in location and time. The analytic solution has been derived from Reference 6.

It will be seen from Figure 4.3.8 that the oscillation of the solution with finite elements in time is much smaller than with the central finite difference method and that the result converges much quicker. The size of the time elements were chosen such that the same number of equations were always solved with a

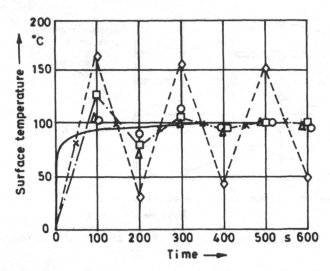

Figure 4.3.8 *Surface temperature of a plane exchanging heat with the environment.* ○ *Cubic time element, initial temperature of plate: 0°C;* △ *square time element, ambient temperature 100°C;* —·—□ *linear time element, half plate thickness 0.05 m;* ———◇ *difference method, conductivity 18 kcal/h m°C;* ——*exact solution, density × specific heat: 912 kcal/m³ °C;* × *averaging technique, heat transfer coefficient 5000 kcal/hm² °C applied to difference method.*

view to getting the result at a definite time. From this aspect no improvement was found using higher order time elements.

In this case, the space discretisation used five parabolic elements (8 nodes per element). A different number of elements (e.g. two parabolic elements only) or a different type of element (e.g. linear space elements) tended to provide the same result.

On studying Figure 4.3.8 it will be recognised that an averaging technique in the difference method stabilises and improves the solution. However, under certain conditions (widely differing time intervals) the solution found by the averaging technique is liable to be less accurate than the original solution.

A similar result was obtained if from time $t = 0$ onwards both plate surfaces were subjected to a constant heat flow per unit area. Figure 4.3.9 illustrates the resulting surface temperature in comparison with the analytic solution derived from Reference 7.

Using the example of a plane plate, the temperature field was determined under conditions of a periodically changing surface temperature. The surface temperature changed in a parabolic manner. This was reproducible in a precise manner by two quadratic or cubic time elements per period. Figure 4.3.10 illustrates the reproduced surface temperature and the temperature at the plate centre under steady-state conditions. This result was reached in practice in the calculation after three time-step periods. After that the variation of the solution was less than

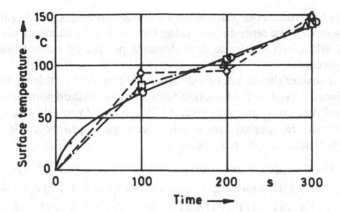

Figure 4.3.9 *Surface temperature of a plane with specified heat flow through the surfaces. ○ Cubic time element, initial temperature of plate: 0°C; △ square time element, material properties: see Figure 4.3.8; —·—□ linear time element, heat flow: 50 000 kcal/h m²; ---◇ difference solution, —— exact solution.*

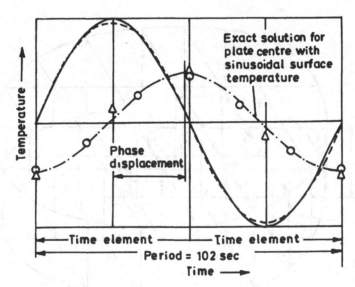

Figure 4.3.10 *Reproduction of a parabolic surface temperature by square and cubic time elements. ○ Cubic time element; △ square time element; ---parabolic surface temperature; ——sinusoidal surface temperature; material properties: see Figure 4.3.8; half plate thickness: 0.02 m.*

1/1000th of the amplitude of the surface temperature. For comparison, the analytic solution for the temperature at the plate centre is given in Figure 4.3.10, this result being for the sinusoidal change of the surface temperature and derived from Reference 8. Since the deviation between the surface temperatures is very small it may be assumed that the reference solution represents the sought solution. It will be seen that the reproduction using two cubic time elements per period is very good, while two quadratic time elements per period do not describe the process so accurately.

Next we consider the case of two-dimensional transient problems, both linear and non-linear, solved by the Galerkin finite element method with elements both in space and time coordinates as given by Bruch and Zyvoloski [9].

The temperature approximation using rectangular prisms as elements and linear shape functions within an element is

$$
\begin{aligned}
T(x, y, t) = &(1/8)(1 + \xi)(1 + \eta)(1 + \zeta)T_i + (1/8)(1 - \xi)(1 + \eta)(1 + \zeta)T_j + (1/8) \\
&\times (1 - \xi)(1 - \eta)(1 + \zeta)T_k + (1/8)(1 + \xi)(1 - \eta)(1 + \zeta)T_l + (1/8)(1 + \xi) \\
&\times (1 + \eta)(1 - \zeta)T_m + (1/8)(1 - \xi)(1 + \eta)(1 - \zeta)T_n + (1/8)(1 - \xi)(1 - \eta) \\
&\times (1 - \zeta)T_o + (1/8)(1 + \xi)(1 - \eta)(1 - \zeta)T_p \\
= &N_i T_i + N_j T_j + N_k T_k + N_l T_l + N_m T_m + N_n T_n + N_o T_o + N_p T_p \quad (4.3.7)
\end{aligned}
$$

where $\xi = 2(x - x_c)/(\Delta x)$; $\eta = 2(y - y_c)/(\Delta y)$; $\zeta = 2(t - t_c)/(\Delta t)$; $x_c, y_c, t_c =$ coordi-

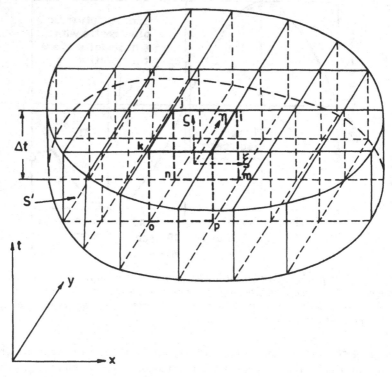

Figure 4.3.11 *Solution domain divided into rectangular prismatic elements in space and time.*

nates of the centroid of each element; T_i, T_j, T_k, T_l, T_m, T_n, T_o, T_p = values of temperature at the appropriate nodal points (Figure 4.3.11); N_i, N_j, N_k, N_l, N_m, N_n, N_o, N_p = shape functions; Δx = x-coordinate spacing; Δy = y-coordinate spacing; Δt = t-coordinate spacing; and i, j, k, l, m, n, o, p = node numbers.

This example is governed by equation (4.2.1) and subject to the following boundary and initial conditions:

$$T(L_x, y, t) = T(x, L_y, t) = 1.0$$

$$\frac{\partial T(0, y, t)}{\partial x} = \frac{\partial T(x, 0, t)}{\partial y} = 0.0 \tag{4.3.8}$$

and

$$T(x, y, 0) = 0.0$$

This example describes the unsteady heat conduction in a long bar of square cross-section. In order to check these results with those of Carnahan *et al.* [10] who used an implicit alternating direction finite difference method, it is assumed that all the constants are 1.0 and the dependent and independent variables are dimensionless in equations (4.3.8). The analytical solution for this problem is

$$T(x, y, t) = 1.0 + \sum\sum C_n \cos \frac{(2n-1)\pi x}{2L_x} \cos \frac{(2j-1)\pi y}{2L_y}$$

$$\times \exp\left[-\left[\frac{k_x(2n-1)^2\pi^2}{4L_x^2} + \frac{k_y(2j-1)^2\pi^2}{4L_y^2}\right]t\right] \tag{4.3.9}$$

where

$$C_n = \frac{16.0(-1.0)(-1)^{n+1}(-1)^{j+1}}{\pi^2(2n-1)(2j-1)} \tag{4.3.9a}$$

Table 4.3.3 lists the analytical results using equations (4.3.9) for the square domain $0 < x < 1.0$, $0 < y < 1.0$, with $k_x = k_y = 1.0$, $\Delta x = \Delta y = 0.1$ and $t = 0.75$.

Table 4.3.3 Dimensionless temperatures at $t = 0.75$ obtained using the analytical solution

0.960	0.960	0.962	0.964	0.968	0.972	0.976	0.982	0.988	0.994	1.000
0.960	0.961	0.962	0.965	0.968	0.972	0.977	0.982	0.988	0.994	1.000
0.962	0.962	0.964	0.966	0.969	0.973	0.978	0.983	0.988	0.994	1.000
0.964	0.965	0.966	0.968	0.971	0.975	0.979	0.984	0.989	0.994	1.000
0.968	0.968	0.969	0.971	0.974	0.977	0.981	0.985	0.990	0.995	1.000
0.972	0.972	0.973	0.975	0.977	0.980	0.983	0.987	0.991	0.996	1.000
0.976	0.977	0.978	0.979	0.981	0.983	0.986	0.989	0.993	0.996	1.000
0.982	0.982	0.983	0.983	0.985	0.987	0.989	0.992	0.994	0.997	1.000
0.988	0.988	0.988	0.989	0.990	0.991	0.993	0.994	0.996	0.998	1.000
0.994	0.994	0.994	0.994	0.995	0.996	0.996	0.997	0.998	0.999	1.000
1.000	1.000	1.000	1.000	1.000	1.000	1.000	1.000	1.000	1.000	1.000

Table 4.3.4 Dimensionless temperatures at $t = 0.75$ obtained using the finite element technique with $\Delta t = 0.05$

0.953	0.953	0.955	0.958	0.962	0.967	0.972	0.979	0.985	0.993	1.000
0.953	0.954	0.956	0.958	0.962	0.967	0.973	0.979	0.986	0.993	1.000
0.955	0.956	0.957	0.960	0.964	0.968	0.974	0.980	0.986	0.993	1.000
0.958	0.958	0.960	0.962	0.966	0.970	0.975	0.981	0.987	0.993	1.000
0.962	0.962	0.964	0.966	0.969	0.973	0.977	0.983	0.988	0.994	1.000
0.967	0.967	0.968	0.970	0.973	0.976	0.980	0.985	0.990	0.995	1.000
0.972	0.973	0.974	0.975	0.977	0.980	0.984	0.987	0.991	0.996	1.000
0.979	0.979	0.980	0.981	0.983	0.985	0.987	0.990	0.993	0.997	1.000
0.985	0.986	0.986	0.986	0.988	0.990	0.991	0.993	0.995	0.998	1.000
0.993	0.993	0.993	0.993	0.994	0.995	0.996	0.997	0.998	0.999	1.000
1.000	1.000	1.000	1.000	1.000	1.000	1.000	1.000	1.000	1.000	1.000

Table 4.3.5 Dimensionless temperatures at $t = 0.75$ obtained by Carnahan *et al.* [10] using an implicit alternating direction finite difference method with $\Delta t = 0.05$

0.960	0.961	0.962	0.964	0.968	0.972	0.976	0.982	0.987	0.994	1.000
0.961	0.961	0.962	0.965	0.968	0.972	0.977	0.982	0.987	0.994	1.000
0.962	0.962	0.964	0.966	0.969	0.973	0.978	0.983	0.988	0.995	1.000
0.964	0.965	0.966	0.968	0.971	0.975	0.979	0.984	0.988	0.995	1.000
0.968	0.968	0.969	0.971	0.974	0.977	0.981	0.986	0.990	0.995	1.000
0.972	0.972	0.973	0.975	0.977	0.980	0.983	0.987	0.991	0.996	1.000
0.976	0.977	0.978	0.979	0.981	0.983	0.986	0.990	0.992	0.997	1.000
0.982	0.982	0.983	0.984	0.986	0.987	0.990	0.992	0.994	0.997	1.000
0.987	0.987	0.988	0.988	0.990	0.991	0.992	0.994	0.996	0.998	1.000
0.994	0.994	0.995	0.995	0.995	0.996	0.997	0.997	0.998	0.999	1.000
1.000	1.000	1.000	1.000	1.000	1.000	1.000	1.000	1.000	1.000	1.000

Table 4.3.4 lists the finite element weighted residual solution for the same problem using a time step $\Delta t = 0.05$. Table 4.3.5 lists the results of Carnahan *et al.* [10] who used an implicit alternating direction finite difference method with the same time step.

The finite element scheme used in the above examples is an implicit scheme and is stable. For the element size used, the results check closely with the analytical results, e.g. in the second example using $\Delta t = 0.05$ h, the maximum deviation from the analytical results was about 2.73%. By decreasing the element size, the finite element solution will converge to the exact solution.

As another example of the application of the technique, the following problem with sharp corners in the boundary is investigated (see Figure 4.3.12). Consider finding $T(x, y, t)$ satisfying the equation

$$\nabla^2 T = \frac{\partial T}{\partial t} \qquad (4.3.10a)$$

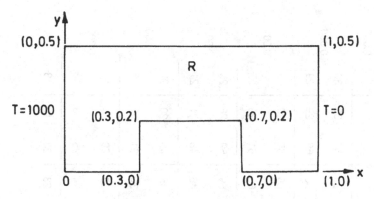

Figure 4.3.12 *Solution domain.*

in the domain R, shown in Figure 4.3.12, with the boundary conditions

$$T(0, y, t) = 1000 \qquad (4.3.10b)$$

$$T(1, y, t) = 0 \qquad (4.3.10c)$$

and $\partial T/\partial n = 0$ on all boundaries, where $\partial/\partial n$ is the derivative normal to the boundary. The initial condition is a small time solution in a plane medium and is taken to be

$$T(x, y, 0) = \text{erfc}\left(\frac{x}{2\sqrt{t}}\right) \qquad (4.3.10d)$$

where $t = 0.0005$ and is equivalent to one time step in the numerical solutions that follow.

Bell [11] presents a method for treating the singularities that occur in the solution to this problem due to the sharp corners in the boundary. His method is essentially an extension of the method due to Motz [12] for solving elliptic problems and approximates to the analytical form of the singularity in terms of neighbouring function values at each time step. His method is used in conjunction with the simple explicit finite difference scheme and subsequently the overall method is explicit.

Bell's results (using five-term approximation, six-term approximation and simple explicit schemes) are given in Table 4.3.6 along with results obtained by using the finite element space–time coordinate method for the x and y spacings and time step shown.

If $k_x = \psi_1 T$ and $k_y = \psi_2 T$ are inserted into equation (4.2.1) it becomes the non-linear partial differential equation

$$\frac{\partial}{\partial x}\left(\psi_1 T \frac{\partial T}{\partial x}\right) + \frac{\partial}{\partial y}\left(\psi_2 T \frac{\partial T}{\partial y}\right) + Q = \rho c \frac{\partial T}{\partial t} \qquad (4.3.11)$$

Table 4.3.6 Solution corresponding to a time of 0.1 s (200 time steps)

1000	842	842	688	687	540	539	404	403	287	286	192	192	120	120	68	68	30	30	0.000
1000	842	842	688	686	540	538	403	403	286	285	192	191	120	119	68	67	30	30	0.000
	846	845	694	692	546	544	408	406	288	287	192	191	119	119	66	66	29	29	
1000	846	845	694	692	545	544	407	406	287	287	191	190	118	117	66	65	29	29	0.00
	856	855	714	711	568	565	418	416	291	290	190	190	112	112	60	61	26	27	
1000	857	856	715	713	566	566	416	417	290	289	190	188	112	111	59	59	26	25	0.000
	874	871	756	749	635	622	427	423	293	292	189	187	88	91	46	48	21	21	
1000	875	873	758	755	635	634	424	425	291	291	188	186	86	86	45	45	20	20	0.000
	891	888	801	795	762	752							38	41	30	31	15	16	
1000	892	890	805	800	767	762							36	36	28	29	14	15	0.000
	898	895	818	813	787	780							29	31	24	25	13	13	
1000	899	897	822	817	791	787							27	28	23	23	12	13	0.000

5-term approx	Simple explicit
6-term approx	Finite element (space–time)

$\Delta x = \Delta y = 0.05$ $\Delta t = 0.0005$

Table 4.3.7 Temperature at $t = 0.2$ h obtained using the finite element technique with $\Delta t = 0.1$ h

1.000	1.000	1.000	1.000	1.000	1.000	1.000	1.000
1.000	0.944	0.889	0.845	0.845	0.888	0.943	1.000
1.000	0.892	0.776	0.674	0.673	0.774	0.891	1.000
1.000	0.853	0.683	0.519	0.516	0.679	0.852	1.000
1.000	0.853	0.682	0.513	0.510	0.677	0.852	1.000
1.000	0.896	0.784	0.677	0.675	0.780	0.896	1.000
1.000	0.950	0.904	0.859	0.858	0.902	0.950	1.000
1.000	1.000	1.000	1.000	1.000	1.000	1.000	1.000

Table 4.3.8 Temperature at $t = 0.4$ h obtained using the finite element technique with $\Delta t = 0.1$ h

1.000	1.000	1.000	1.000	1.000	1.000	1.000	1.000
1.000	0.988	0.979	0.973	0.973	0.979	0.988	1.000
1.000	0.980	0.964	0.954	0.954	0.964	0.980	1.000
1.000	0.976	0.957	0.943	0.943	0.956	0.976	1.000
1.000	0.976	0.958	0.944	0.944	0.958	0.976	1.000
1.000	0.982	0.968	0.957	0.957	0.968	0.982	1.000
1.000	0.991	0.984	0.978	0.978	0.984	0.991	1.000
1.000	1.000	1.000	1.000	1.000	1.000	1.000	1.000

where ψ_1 and ψ_2 are assumed to be constants. Take $\psi_1 = \psi_2 = 1.0$, $Q = 0$, $\rho c = 1.0$, and the following auxiliary conditions:

$$T(0, y, t) = T(x, 0, t) = T(L_x, y, t) = T(x, L_y, t) = 1.0 \qquad (4.3.11a)$$

and

$$T(x, y, 0) = 0.1 \qquad (4.3.11b)$$

The system of non-linear algebraic equations that results from the application of the finite element scheme was solved by Powell [13]. The results are presented in Tables 4.3.7 and 4.3.8 for $t = 0.2$ h and $t = 0.4$ h, respectively.

Many more examples of unsteady heat transfer are available in the literature [14–21] which further illustrate the application of finite element analysis to practical problems.

REFERENCES

[1] Carslaw H. S. and Jaeger J. C., *Conduction of Heat in Solids* Clarendon Press, Oxford (1954).

[2] Wilson E. L. and Nickell R. E., Application of the finite element method to heat conduction analysis *Nucl. Engg. Des.*, **4**, 276–286 (1966).

[3] Zienkiewicz O. C. and Parekh C. J., Transient field problems: two-dimensional analysis by isoparametric finite elements *Int. J. Num. Meth. Engg*, **2**, 61–71 (1970).

[4] Donea J., On the accuracy of finite element solution to the transient heat conduction equation, *Int. J. Num. Meth. Engg*, **8**, 103–110 (1974).

[5] Kohler W. and Pitter J., Calculation of transient temperature fields with finite elements in space and time dimensions, *Int. J. Num. Meth. Engg*, **8**, 625–631 (1974).

[6] VDI-Warmeatlas, VDI Verlag, Dusseldorf (1957).

[7] Bachr H. D., Die Losung nichtstationarer warmeleitungsprobleme mit Hilfe der Laplace—Transformation *Forschung auf dem Gebiete des Ingenieurwesons*, 21, Heft 2 Dusseldorf (1955)

[8] Grober H., Erka S. and Grigull V., *Grund geretzeder Warmebertragung* Springer-Verlag, Berlin (1963).

[9] Bruch J. C. and Zyvoloski G., Transient two-dimensional heat conduction problems solved by the finite element method *Int. J. Num. Meth. Engg*, **8**, 481–494 (1974).

[10] Carnahan B., Luther H. A. and Wilkes J. O., *Applied Numerical Methods* John Wiley & Sons, New York, pp. 454–461 (1969).

[11] Bell G. E., *A Method for Treating Boundary Singularities in Time Dependent Problems* TR18, Dept of Math., Brunel Univ., Uxbridge, Middlesex, 19 pp. (1972)

[12] Motz, H., The treatment of singularities of partial differential equations by relaxation methods *Quart. J. Appl. Math.*, **4**, 371–377 (1946).

[13] Powell M. J. D., A Fortran subroutine for solving systems of nonlinear algebraic equations, *Numerical Methods for Nonlinear Algebraic Equations*, Ed. Rabinowitz Gordon and Breach, New York, pp. 115–161 (1970).

[14] Lewis R. W. and Bass B. R., The determination of temperature and stresses in cooling bodies by finite elements *ASME Trans. J. Heat Transfer*, **98**, 478–484 (1976).

[15] Nayak, G. C. and Dua S. N. Thermal analysis of steam turbine casings under steady state and transients, *Proc. I. Int. Conf. on Num. Meth. in Thermal Problems* Swansea, pp. 824–839 (1979).

[16] Thomas, B. G., Samaria Sekera I. V. and Brimacombe J. K., Comparison of numerical modelling techniques for complex, two-dimensional, transient heat conduction problems *Metall. Trans.* B, **15B**, 307–318 (1984).

[17] Czomber L., Haberland C. and Lahrmann, A., A new stable finite element formulation for the determination of transient temperatures in arbitrarily shaped structures *Numerical Methods in Thermal Problems*, Vol. IV, Part 1, pp. 5–17 Eds. R. W. Lewis and K. Morgan Pineridge Press (1985).

[18] Dahlblom Ola, A finite element program for simulation of temperature and hardening in concrete structures *Numerical Methods in Thermal Problems*, Vol. VI, Part 1, pp. 171–181, Pineridge Press (1989).

[19] Vila Real P. M. M. and Magalhaes Oliverna C. A., Use of polynomial functions for hierarchical improvement of the finite element solution for transient heat conduction problems with higher thermal gradients *Numerical Methods in Thermal Problems*, Vol. VI, Part 1, pp. 227–240, Eds. R. W. Lewis and K. Morgan, Pineridge Press (1989).

[20] Hughes T. J. R., Unconditionally stable algorithms for nonlinear heat conduction *Comp. Math. Appl. Mech. Engg*, **10**, 135–139 (1977).

[21] Cha'o-Kuang Chen and Tzer-Ming chen, New hybrid Laplace transform/finite element method for three-dimensional transient heat conduction problem *Int. J. Num. Meth. Engg*, **32**, 45–61 (1991).

5 Phase Change Problems— Solidification and Melting

In this chapter we will consider an important class of problems which come under the category of phase change. Solidification and melting are important processes in many engineering fields. The chapter begins with an introduction to phase change problems and their importance in industry. The mathematical formulation of phase change problems is then introduced. Then follows a description of several numerical methods used in modelling the phase change problems. A bench-mark example is considered for a one-dimensional solidification and melting process. Application of the theory to some industrial problems is then dealt with.

5.1 INTRODUCTION

Materials processing, metallurgy, purification of metals, growth of pure crystals from melts and solutions, solidification of casting and ingots, welding, electroslag melting, zone melting, thermal energy storage using phase change materials, etc. involve melting and solidification. These phase change processes are accompanied by either absorption or release of thermal energy. A moving boundary exists that separates the two phases of differing thermo-physical properties and at which thermal energy is either absorbed or liberated. If we consider the solidification of a casting or ingot, the superheat in the melt and the latent heat liberated at the solid–liquid interface are transferred across the solidified metal, interface and the mould, encountering at each of these stages a certain thermal barrier. In addition, the metal shrinks as it solidifies and an air gap is formed. Thus, additional thermal resistance is encountered. The heat transfer processes occurring are complex. The cooling rates employed range from 10^{-5} to 10^{10} K s^{-1} the corresponding solidification systems extend from depths of several metres to a few micrometres. These various cooling rates produce different micro structures and hence a variety of thermomechanical properties. During the solidification of binary and multi-component alloys, the physical phenomena become more complicated due to phase transformation taking place over a range of temperatures. During the solidification of an alloy the concentrations vary locally from the original mixture, as material may be preferentially incorporated or rejected at the solidification front. The material between the solidus and the liquidus temperatures is partly solid and partly liquid and resembles a porous medium and is referred as a 'mushy zone'.

A complete understanding of the phase change phenomenon involves the analysis of the various processes that accompany it. The most important of these processes, from a macroscopic point of view, is the heat transfer process. This is complicated by the release or absorption of the latent heat of fusion at the solid–liquid interface. Several methods have been used to take into account the liberation of latent heat. These methods are generally divided into fixed and moving mesh methods. Fixed mesh methods involve the solution of a continuous system with an implicit representation of the phase change, while in the 'front tracking' or 'moving mesh' methods, the solid and liquid regions are treated separately and the phase change interface is explicitly determined as a moving boundary. A comprehensive treatment of the subject of moving boundary problems in general and phase change related moving boundary problems in particular appears in a book by Crank [1]. A useful review of the methods used in a finite element context can be found in a paper by Dalhuijsen and Segal [2]. Salcudean and Abdullah [3] have listed several latent heat formulations and many references for general numerical modelling of phase change. Recently Voller *et al.* [4] have discussed the fixed grid methods with the aim of providing a comprehensive review of the major formulations with an emphasis on the numerical features.

5.2 STATEMENT OF THE PROBLEM

The classical problem involves considering the conservation of energy in the domain by dividing this into two distinct domains, Ω_ℓ and Ω_s, where $\Omega_\ell + \Omega_s = \Omega$. The energy conservation is written for the one-dimensional case, for simplicity, as

$$\rho_\ell c_\ell \frac{\partial T}{\partial t} = K_\ell \frac{\partial^2 T}{\partial x^2} \quad \text{in } \Omega_\ell \tag{5.2.1}$$

and

$$\rho' c_s \frac{\partial T}{\partial t} = K_s \frac{\partial^2 T}{\partial x^2} \quad \text{in } \Omega_s \tag{5.2.2}$$

where the subscripts ℓ and s denote liquid and solid, respectively. The complete description of the problem involves, in addition to the initial conditions and the appropriate external boundary conditions, the interface conditions on the phase change boundary, $\Gamma_{s\ell}$, which are

$$T_{s\ell} = T_f$$

$$-K_s \left(\frac{\partial T}{\partial x}\right)_s = \rho_s L \frac{ds}{dt} - K_\ell \left(\frac{\partial T}{dx}\right)_\ell \quad \text{on } \Gamma_{s\ell} \tag{5.2.3}$$

where s represents the position of the interface, $\mathrm{d}s/\mathrm{d}t$ the interface velocity and T_f is the phase change temperature. Equation (5.2.3) states that the heat transferred by conduction in the solidified portion is equal to the heat entering the interface by latent heat liberation at the interface and the heat coming from the liquid by conduction. The main complication in solving the classical problem lies in tracking interface boundary positions.

5.3 NUMERICAL METHODS FOR MODELLING PHASE CHANGE

The fixed grid methods offer a more general solution since they account for the phase change conditions implicitly without attempting *a priori* to establish the position of the front. These methods are based on a weak formulation of the classical problem, which is commonly referred to as the enthalpy formulation. A single energy conservation equation is written for the whole domain as

$$\frac{\partial H}{\partial t} = K \frac{\partial^2 T}{\partial x^2} \quad \text{in } \Omega \tag{5.3.1}$$

where H is the enthalpy function or the total heat content which is defined for isothermal phase change as

$$H(T) = \int_{T_r}^{T} \rho c_s(T)\,\mathrm{d}T \qquad (T \leqslant T_f)$$

$$H(T) = \int_{T_r}^{T_f} \rho c_s(T)\,\mathrm{d}T + \rho L + \int_{T_f}^{T} \rho c_\ell(T)\,\mathrm{d}T \quad (T \geqslant T_\ell) \tag{5.3.2}$$

and for phase change over an interval of temperatures T_s to T_ℓ, which are the solidus and the liquidus, respectively, we have

$$H(T) = \int_{T_r}^{T_s} \rho c_s(T)\,\mathrm{d}T + \int_{T_s}^{T} \left[\rho\left(\frac{\mathrm{d}L}{\mathrm{d}T}\right) + \rho c_f(T) \right]\mathrm{d}T \qquad (T_s < T \leqslant T_\ell) \tag{5.3.3}$$

$$H(T) = \int_{T_r}^{T_s} \rho c_s(T)\,\mathrm{d}T + \rho L + \int_{T_s}^{T_\ell} \rho c_f(T)\,\mathrm{d}t + \int_{T_\ell}^{T} \rho c_\ell(T)\,\mathrm{d}T \quad (T \geqslant T_\ell)$$

where c_f is the specific heat in the freezing interval, L is the latent heat and T_r is a reference temperature below T_s.

Among the fixed mesh methods, one of the earliest and the most commonly used methods has been the 'effective heat capacity' method. This method is derived from writing

$$\frac{\partial H}{\partial t} = \frac{\partial H}{\partial T}\frac{\partial T}{\partial t} = K \frac{\partial^2 T}{\partial x^2} \quad \text{in } \Omega \tag{5.3.4}$$

Compared with the standard heat conduction equations such as (5.2.1) and (5.2.2), one can write

$$c_{eff} = dH/dT \qquad (5.3.5)$$

where c_{eff} is the effective heat capacity. This can be evaluated directly from equation (5.3.3):

$$c_{eff} = \rho c_s \qquad (T < T_s)$$

$$c_{eff} = \rho c_f + \frac{L}{T_\ell - T_s} \qquad (T_s < T < T_\ell) \qquad (5.3.6)$$

$$c_{eff} = \rho c_\ell \qquad (T > T_\ell)$$

Figure 5.3.1 shows a typical variation of the effective heat capacity and enthalpy with temperature. It can be seen from the figure that if this directly evaluated effective specific heat is used, it will be necessary to maintain an interval of temperature for the evolution of latent heat, otherwise the effective heat capacity will become infinite. This method therefore cannot accurately model an isothermal phase change due to the requirement of a temperature range. Due to the step-like behaviour of c_{eff} around the phase change interval, numerical oscillations may occur, making the achievement of a convergent solution difficult.

In order to overcome the difficulties encountered in using a directly evaluated effective capacity, recourse is made to several averaging techniques which are

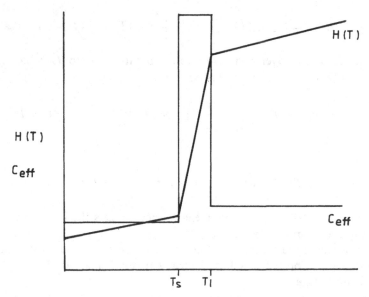

Figure 5.3.1 *Typical variation of enthalpy and c_{eff} with temperature.*

generally referred to as the enthalpy method. The enthalpy method enables the heat capacity to be defined as a smooth function of temperature. The simplest approximation used for averaging, for the two-dimensional case, is

$$\frac{dH}{dT} = \frac{1}{2}\left[\frac{\frac{\partial H}{\partial x}}{\frac{\partial T}{\partial x}} + \frac{\frac{\partial H}{\partial y}}{\frac{\partial T}{\partial y}}\right] \tag{5.3.7}$$

This method has been reported to cause oscillations in certain circumstances. A better method, as reported by Del Guidice et al. [5], is

$$\frac{dH}{dT} = \left[\frac{\frac{\partial H}{\partial x}\frac{\partial T}{\partial x} + \frac{\partial H}{\partial y}\frac{\partial T}{\partial y}}{\left(\frac{\partial T}{\partial x}\right)^2 + \left(\frac{\partial T}{\partial y}\right)^2}\right] \tag{5.3.8}$$

Lemmon [6] suggests another approximation which is reported to work satisfactorily:

$$\frac{dH}{dT} = \left[\frac{\left(\frac{dH}{dx}\right)^2 + \left(\frac{dH}{dy}\right)^2}{\left(\frac{\partial T}{\partial x}\right)^2 + \left(\frac{\partial T}{\partial y}\right)^2}\right]^{1/2} \tag{5.3.9}$$

Morgan et al. [7] have advocated the use of a simple backward difference approximation

$$\frac{dH}{dT_n} = \left(\frac{H_n - H_{n-1}}{T_n - T_{n-1}}\right) \tag{5.3.10}$$

where n represents the time step number. Lewis and Roberts [8] claim that the last scheme is computationally quicker than the other averaging techniques.

In using the above techniques in a finite element analysis it is common practice to interpolate H from the nodal values using the same basis functions as for T, thus obtaining a smoothing effect.

A fixed mesh method which is gaining favour and is essentially an enthalpy based method as well, is the 'Fictitious Heat-Flow Method' first suggested in the finite element context by Ralph and Bathe [9]. This is also referred to as the 'heat source method'. The central idea behind the heat source method is to lump all the latent heat available at the nodes, which may then be released or absorbed as an internal heat source at the appropriate temperature or a range of temperatures. The implementation of this method in a finite element heat transfer program is not straightforward. Ralph and Bathe [9] have given considerable details for such an implementation.

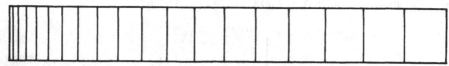

$$T = -45.0$$

| $T_o = 0.0$ | $T_f = -0.15$ |

$$x = 1.0 \qquad x = 4.0$$

$$L = 70.26$$
$$\rho c = 1.0$$
$$k = 1.0$$

Figure 5.4.1 *Solidification example.*

No. of Nodes: 123   No. of Elements : 20

No. of Nodes/Element : 9

Figure 5.4.2 *Finite element mesh for bench-mark example.*

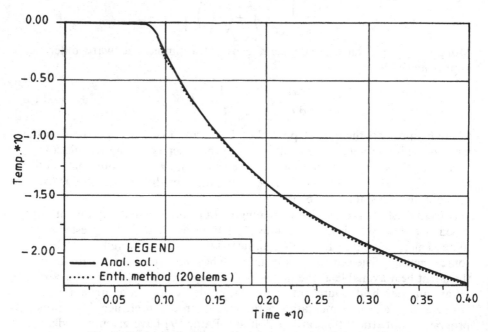

Figure 5.4.3 *Enthalpy method versus the analytical method.*

5.4   BENCH-MARK EXAMPLE

A one-dimensional bench-mark example of solidification as shown in Figure 5.4.1 is solved and compared with the analytical solution. No particular units are necessary. This example has been used for such comparisons by several investigators [7, 9]. The temperature versus time curve at a selected location ($x = 1.0$ here) in the domain is used for the comparisons. The mesh used to solve the above example is shown in Figure 5.4.2. Figure 5.4.3 shows a good comparison between the enthalpy and analytical solution [11] for the bench-mark problem.

An example of phase change over a range is solved using all the three methods mentioned earlier, which does not have an analytical solution. The problem description is shown in Figure 5.4.4. The results from all the three methods are shown in Figure 5.4.5. The enthalpy and source methods approximately agree. The

$$T = -45.0 \quad \boxed{T_0 = 0.0 \qquad T_s = -10.15 \quad T_l = -0.15} \qquad L = 70.26$$

$$\rho c = 1.0$$

$$x = 1.0 \qquad\qquad\qquad z = 4.0 \qquad k = 1.0$$

Figure 5.4.4   *Example of solidification over a temperature range.*

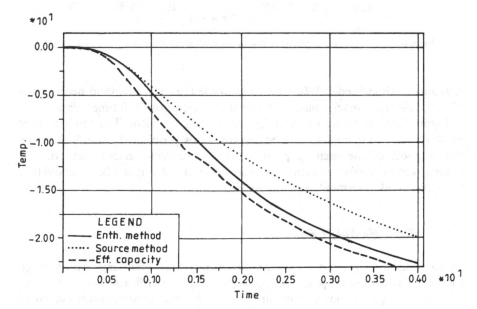

Figure 5.4.5   *Solidification over a temperature range ($T_l - T_s = 10.0$).*

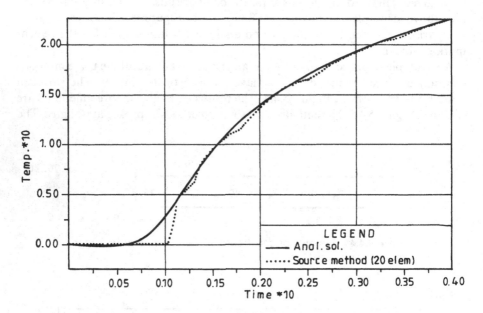

Figure 5.4.6   *Melting example.*

Figure 5.4.7   *Source method is the analytical solution for melting example.*

enthalpy method needed 1500 time steps, while the source method needed only 70. The effective capacity method result is obtained from 120 time steps.

Figure 5.4.6 shows an example for the melting problem. The finite element mesh shown in Figure 5.4.2 is used again for this problem. Figure 5.4.7 shows the solution of the melting problem using the heat source method. For comparison the analytical solution is also shown in 5.4.7. It can be observed that there is a good agreement between the two.

5.5   EXAMPLES

In this section we present some examples of problems involving castings since these form an important practical application of the solidification process. Cases of ingot castings without a mould and then with a mould are considered, where enough experimental data are available for comparison. The extension of the method to the case of continuous casting is considered next. The method can be extended to diecastings, centrifugal castings, squeeze castings, etc.

The study of the process of heating/cooling bodies, allowing for latent heat of solidification and phase transformation, is a complex problem which cannot always be treated in a closed form. Even the limited available closed-form solutions are for semi-infinite domains. When the finite dimensions and true shape of the body are taken into account, the problem is further complicated. The solidification of steel ingots is a typical example of this type of study.

An experimental determination of the temperature and thermal state is not only complex and laborious but also cannot be properly carried out. Analytical methods of calculation often provide results that do not coincide with practical data owing to a number of assumptions and simplifications used in the analysis such as one-dimensional heat conduction, semi-infinite geometry, constant properties, etc.

Although the solidification of an ingot is a three-dimensional process it is possible to simulate the process by using a two-dimensional model [15]. It has been demonstrated that for the ingot dimensions used in actual practice a true picture of the solidification process can be obtained by calculation in a horizontal section at mid height through the ingot. Moreover, if the ratio between the width and thickness of the ingot is greater than 1.8, accurate calculations can also be performed in a vertical section across the narrow side. In some instances, ingots may be required for rolling purposes, and in such a case reheating of the ingot is undertaken. Thermal economy dictates that the heating costs should be kept to a minimum.

A finite element method of solution for heat transmission in steel ingots was made by Saliman and Fakhroo [14] who used a variational formulation and triangular elements to describe the two-dimensional problem of heat conduction in ingots. They also accounted for the variation of specific heat and thermal conductivity.

5.5.1 Ingot casting

In the present analysis the unified approach of Comini *et al.* [15] is used, which permits the simultaneous temperature dependence of thermal conductivity, heat capacity, rate of heat generation and surface heat transfer coefficients.

The governing equation is the two-dimensional transient heat conduction equation given by

$$\rho c \frac{\partial T}{\partial t} = \frac{\partial}{\partial x}\left[K_x \frac{\partial T}{\partial x} \right] + \frac{\partial}{\partial y}\left[K_y \frac{\partial T}{\partial y} \right] + Q \tag{5.5.1}$$

where K_x and K_y are temperature dependent conductivity coefficients, ρc is the temperature-dependent heat capacity and Q is the rate of heat generation. The boundary conditions encountered in ingot castings are

$$K\frac{\partial T}{\partial \eta} = q + \alpha(T - T_x) \qquad \text{(heat flux or convection)} \qquad (5.5.2)$$

and/or

$$K\frac{\partial T}{\partial \eta} = \varepsilon\sigma(T^4 - T_x^4) = 0 \qquad \text{(radiation)} \qquad (5.5.3)$$

Heat transfer by radiation is expressed as

$$q_1 = \varepsilon\sigma(T^4 - T_x^4) = h_r(T - T_x) \qquad (5.5.4)$$

In general K, ρc, Q and α_r are temperature-dependent functions, the last variation being expressed by equation (5.5.4) and others being known numerically. The spacewise discretisation of equation (5.5.1) subject to boundary conditions (5.5.2) and (5.5.3) can be accomplished using Galerkin's method. The region of interest, Ω, is divided into a number of 8-noded isoparametric elements, Ω^e, with quadratic shape functions, N_i, associated with each node i. The unknown function, T, is approximated throughout the solution domain at any time t by

$$T = \sum_{i=1}^{n} N_i(x, y)T_i(t) \qquad (5.5.5)$$

where $T(t)$ are the nodal parameters. The substitution of expression (5.5.5) into (5.5.1) and the application of the Galerkin method results in the following equation:

$$\mathbf{M}\frac{d\mathbf{T}}{dt} + \mathbf{KT} + \mathbf{F} = 0 \qquad (5.5.6)$$

Typical matrix elements are

$$K_{ij} = \sum_{\Omega^e} \left(K_x\frac{\partial N_i\partial N_j}{\partial x\partial x} + K_y\frac{\partial N_i}{\partial y}\frac{N_j}{\partial y} \right)d\Omega + \int_{\Gamma^e} (\alpha + \alpha_r)N_iN_j d\Gamma \qquad (5.5.7)$$

$$M_{ij} = \sum \int_{\Omega^e} \rho C N_iN_j d\Omega \qquad (5.5.8)$$

$$F_i = -\sum \int_{\Omega^e} N_iQ\,d\Omega + \sum \int_{\Gamma} N_i(q - \alpha T_x - \alpha_r T_x)d\Gamma \qquad (5.5.9)$$

In the above, the summations are taken over the contributions of each element, Ω^e is the element region and Γ^e refers only to the elements with external boundaries on which conditions (5.5.2) or (5.5.3) are specified. The evaluation of the integral in equation (5.5.8) is based on the interpolation of enthalpy, H, instead of the heat capacity since enthalpy is a smooth function of temperature even in the phase change region. Thus, it is reasonable to interpolate enthalpy, and hence the relationship

$$H = \sum_{i=1}^{n} N_i(x, y)H_i(t) \tag{5.5.10}$$

is used. A simple and more accurate determination of ρc, as shown by Morgan et al. [7], is used in the calculation:

$$\rho c = (H^m - H^{m-1})/(T^m - T^{m-1}) \tag{5.5.11}$$

To avoid iteration within each time step, the unconditionally stable, three-level scheme proposed by Lees (16) is used here:

$$T_{m+1} = -\left(K_m + \frac{3}{2\Delta t}M_m\right)^{-1}\left(K_m T_m + K_m T_{m-1} - \frac{2}{3\Delta t}M_m T_{m-1} + 3F_m\right) \tag{5.5.12}$$

A correct estimate of the radiation coefficient, α_r, in the time interval $t - \Delta t$ to $t + \Delta t$, as suggested by Comini et al. [15], is used here (when $T_\infty = $ constant):

$$\alpha_r = \sigma\varepsilon[(T^2 + T_t^2)(T_t + T_\infty)] + [(T_t - T_{t-\Delta t})^2(T_t + (1/3)T_\infty)] \tag{5.5.13}$$

The solution of the system of simultaneous equations is carried out using a band solver.

Weingart [17] has measured the thickness of the solidified shell in ingots of about 6 tons with dimensions $2100 \times 685 \times 615\ mm^3$. The results, along with the calculated curve by the present method, are shown in Figure 5.5.1. In the

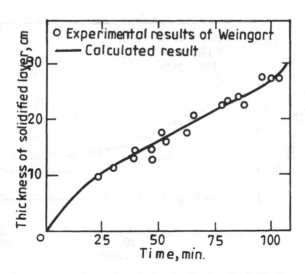

Figure 5.5.1 *Comparison of experimental and calculated solidified layer temperatures.*

calculations a higher heat transfer coefficient (based on purely radiative heat transfer between ingot and mould) is used for the mid-sides compared with the corner. The calculated shell thickness agrees well with the experimental results. The results from the present model are also compared with the measurements of surface temperature and solidification time made on a 1.4 ton ingot by den Hartog *et al.* [18]. Figure 5.5.2 shows the variation of heat transfer coefficients with time measured on the perimeter of a 1.4 ton ingot. These variations of heat transfer coefficient on the perimeter with time are taken into account in the present model. Figure 5.5.3 shows the comparison of surface temperature as a function of time for the 1.4 ton ingot. Agreement between the calculated and the measured value is good. The calculated solidification time is 27 min, which compares well with the experimentally determined value of 28 min [19].

A height of 1.8 m, a size frequently used in steelworks, is adopted for the study of ingots of constant height. Three ingot thicknesses are selected (42, 60, and 84 cm) corresponding to ingots of 2.5, 5 and 10 tons (in the case of square cross-sections). The width is chosen to have width-to-thickness ratios of 1, 1.5, 2 and 3. It is assumed that the powder used on top of the ingot to reduce heat losses is a perfect insulator and that the heat transfer from the base to the ground is such that the temperature of the lower face of the base remains constant at 100 °C.

The solidification times in the middle section of the various ingots mentioned earlier are determined and listed in Table 5.5.1. The ratios f_1 are in close agreement with Sevrin's value [13]. Figure 5.5.4 shows the variation of f_1 with width-to-thickness ratio of the ingots at the same relative distance from the surface of the ingot. It can be observed that the ratio f_1 does not depend on the

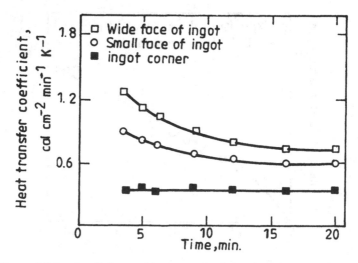

Figure 5.5.2 *Coefficients of heat transfer behaviour: ingot and mould.*

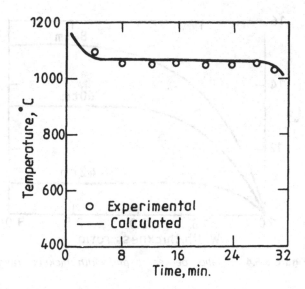

Figure 5.5.3 *Comparison of experimental and calculated ingot temperatures.*

Table 5.5.1 Solidification time at mid-height

Thickness (cm)	Width (cm)	Width thickness	Solidification time (min)	f_1 (average values in parentheses)
42	42	1	45	1
60	60	1	83	1
84	84	1	150	1
				(1)
42	63	1.5	60	1.33
60	90	1.5	111	1.34
84	126	1.5	204	1.36
				(1.343)
42	84	2.0	66	1.46
60	120	2.0	124	1.49
84	168	2.0	225	1.50
				(1.483)
42	126	3.0	69	1.53
60	180	3.0	129	1.55
84	252	3.0	237	1.58
				(1.553)

width of the ingot, but merely on the width-to-thickness ratio. The total solidi-fication times, calculated in minutes, agree well with the formula given by Sevrin [13].

The present method permits the calculation of the solidification time at various levels. The solidification times at different heights of three square ingots are

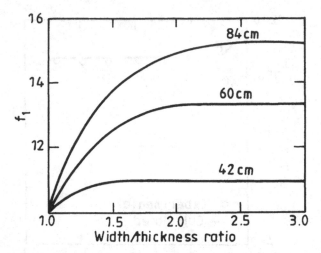

Figure 5.5.4 *Relation between f_1 and width thickness ratio.*

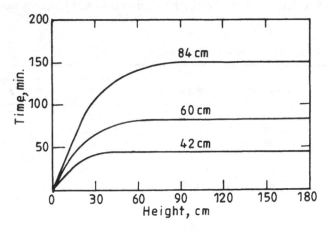

Figure 5.5.5 *Vertical solidification of square ingots.*

shown in Figure 5.5.5. At the base of the ingot, the solidification rate is affected only by the heat flow towards the stool; as the distance from the ingot base increases, the extent of cooling through the walls of the mould increases and the longitudinal heat flow towards the base of the ingot decreases. Since it has been assumed that there is no heat loss from the top of the ingot, all the curves are flat after some time. If the heat loss from the top of the ingot is considered, the drooping characteristic near the top will be observed. It is also seen that the region of the influence of the base cooling on the total solidification time is limited to a height slightly smaller than the thickness of the ingot in the case of square ingots.

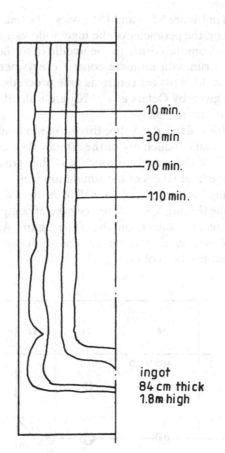

10 min.
30 min
70 min.
110 min.

ingot
84 cm thick
1.8 m high

Figure 5.5.6 *Solidification fronts at mid-plane of narrow section.*

Figure 5.5.6 shows the solidification fronts on the narrow vertical mid-plane at four different times during cooling in an ingot of thickness 84 cm and height 1.8 m.

5.5.2 Ingot casting with mould

Because of the complexities and practical difficulties of experiments on ingot castings, only a limited number of experimental investigations are reported in sufficient detail in the literature to make possible a comparison with the results of a mathematical model. Oeters *et al.* [20] have reported extensive results of the investigation on a 6 ton ingot casting, and this is the example chosen for illustration [32]. The finite element mesh used in the analysis consists of 52 elements (illustrated in Figure 5.5.7) and 185 nodes. The time of air gap formation at various locations on the perimeter of the ingot follows approximately a parabolic law proceeding from the corner to the middle of the face (see Figure 5.5.8). Initially the model is run with intimate contact everywhere, and then the gap opens from the corner towards the centre as time proceeds. In the gap, the heat transfer coefficients given by Oeters et al. [20] are utilised as functions of time in the form shown in Figure 5.5.9.

Since the gap width is fixed at 0.5 mm, this variation is supplied to the model as a variation in thermal conductivity of the air gap. At the start, the time steps need to be very small, of the order of seconds, and they gradually increase up to a value of 2 min in the later stages of the simulation run.

The calculated temperature of the inner wall of the mould for various locations is plotted against time (Figure 5.5.10). For the sake of comparison, the results of Oeters et al. [20] are also shown on the same figure. As the exact thermal properties used by Oeters et al. [20] are not known, it can be concluded that a reasonable agreement has been obtained.

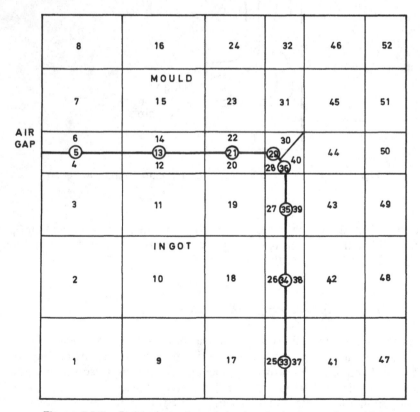

Figure 5.5.7 *Finite element mesh for ingot, air gap and mould.*

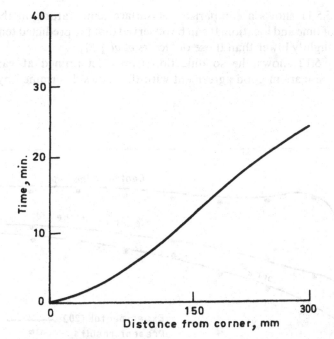

Figure 5.5.8 *Time for air gap formation.*

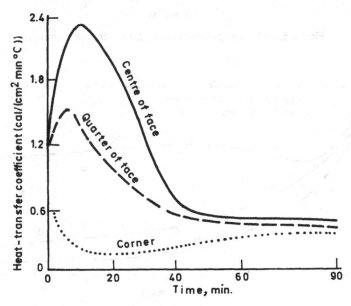

Figure 5.5.9 *Average heat transfer coefficients in air gap.*

Figure 5.5.11 shows a comparison of surface temperature on the ingot as functions of time and location. It can be observed that the predicted temperatures are again slightly lower than those of Oeters *et al.* [20].

Figure 5.5.12 shows the solidification fronts determined at various time intervals; these are in good agreement with the values determined by Weingert [17].

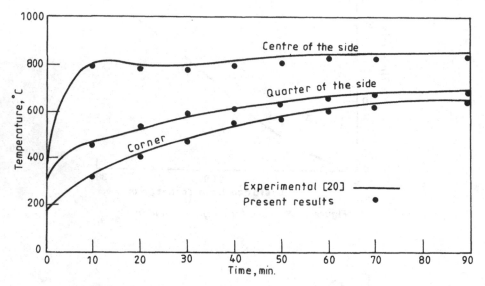

Figure 5.5.10   *Temperature curve for inner wall of mould.*

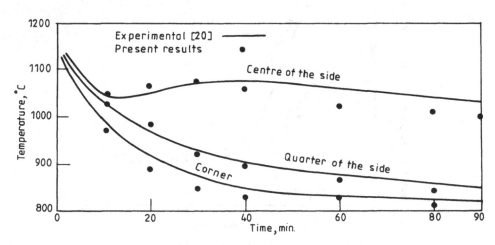

Figure 5.5.11   *Average values of ingot surface temperature.*

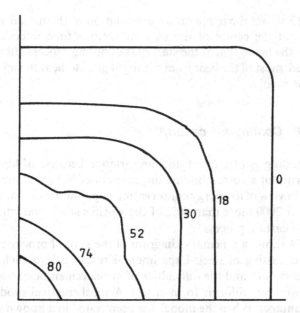

Figure 5.5.12   *Solidification fronts for one quadrant of the ingot (numbers denote time in minutes).*

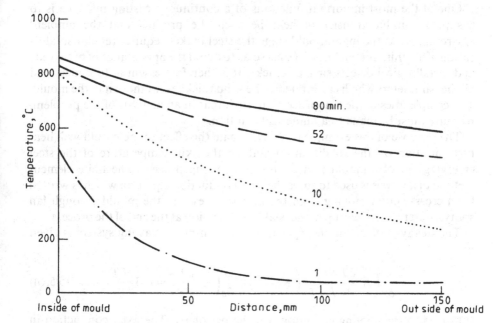

Figure 5.5.13   *Temperature distribution in the mould at the centre of one side at various times after the pour.*

Figure 5.5.13 shows the temperature distribution in the mould wall, which is 150 mm thick, at the centre of the side, at various time intervals. It can be observed from the figure that at the start of the cooling process after pouring has been completed, most of the heat from the ingot goes to heat the mould and very little leaves the mould.

5.5.3 Continuous casting

Continuous casting is of interest and importance because of higher yields of quality steel without a corresponding increase either in capital input or energy. It is expected, in view of its energy conservation capability and lower capital cost, that by the year 2000 more than 60% of the world's steel production will be by the continuous casting process.

Figure 5.5.14 shows a schematic diagram of the several processes involved in the continuous casting of steel. Experimental research into such a process is expensive to carry out and the difficulties of measurement are so great that the results are sometimes difficult to interpret. A mathematical model, therefore, often has advantages. When the model has been tested in a known situation and is seen to describe correctly the effects of most of the important parameters, it can be used to examine the influence of design and operational variables in a real process.

One of the most important functions of a continuous casting machine is to ensure a suitable transfer of heat between the product and the ambient environment. At the ingot mould stage the steel maker requires regular solidification of a minimum thickness of steel so as to avoid the appearance of breakouts and metallurgical defects such as cracks. It is therefore essential to understand all the parameters which can influence the solid thickness at the exit of the mould. The complexities of the real process are such that in an analysis of the problem, recourse must be made to a numerical solution.

The purpose of this example is to investigate the effect of the mould wall heat flux variation on the solidification and on the skin temperature of the steel emerging out of a mould in a continuous casting process. The finite element method of analysis is used to solve the heat conduction equation which is written for a cross-section moving from the meniscus level in the mould through fan sprays, water sprays, etc. up to the beaking off section at the end of the process.

The steady-state equation applicable to a continuous casting system is given by

$$\frac{\partial}{\partial x}\left(K_y\frac{\partial T}{\partial x}\right) + \frac{\partial}{\partial y}\left(K_y\frac{\partial T}{\partial y}\right) + \frac{\partial}{\partial z}\left(K_z\frac{\partial T}{\partial z}\right) = \rho c W\frac{\partial T}{\partial z} \qquad (5.5.14)$$

subjected to the cooling conditions on the periphery. The axial conduction in the direction in which the metal moves is neglected. The movement of a given

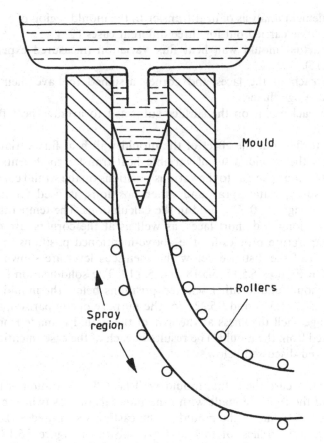

Figure 5.5.14 *Schematic diagram of a continuous caster.*

cross-section in the direction of motion, i.e. z, is given by Wt in the time interval t, where W is the speed of strand which is constant. Thus the right-hand side of equation (5.5.14) is given by

$$\rho c W \frac{\partial T}{\partial (W t)} = \rho c \frac{\partial T}{\partial t} \text{ since } W = \text{constant}$$

Thus the governing equation for the continuous casting system is given as

$$\frac{\partial}{\partial x}\left(K_x \frac{\partial T}{\partial x} \right) + \frac{\partial}{\partial y}\left(K_y \frac{\partial T}{\partial y} \right) = \rho c \frac{\partial T}{\partial t}$$

which is the same as equation (5.5.1) except for the generation term. The boundary and initial conditions are similar to the earlier ingot problem except that the cross-section is considered to move along the strand with time.

A finite element analysis of heat transfer in the mould region of a continuous casting system is carried out for three different cases [21]:

(i) With actual mould wall heat flux variation (measured experimentally) (Figure 5.5.15).

(ii) With each of the faces and corner assumed to have their respective individual average fluxes.

(iii) With each point on the surface subjected to average heat flux for the entire mould.

In order to find out the effect of the mould wall heat flux variation on the strand below the mould, a set of variable heat transfer coefficients (shown in Figure 5.5.16), as applicable to the various positions on the strand corresponding to the fan spray, water spray and other positions, is used for all cases. A temperature range of 30 °C is used in the calculations. The temperatures at the middle of the long and short faces, as well as at the corners, are determined and the temperature profiles for the above-mentioned positions in the strand as a function of the distance below the meniscus level are shown for all the three cases in Figures 5.5.17, 5.5.18 and 5.5.19. The solidification fronts at the exit of the mould and at other selected positions below the mould are shown in Figures 5.5.20, 5.5.21 and 5.5.22. For the purpose of comparison, an estimate of the average shell thickness at the exit of the mould is made from the total heat removed from the mould. The results of each of the cases mentioned above are shown and discussed below.

Case 1 In this case the actual mould wall heat flux variation is used in the analysis and the runs are made with time intervals of 1.2 s (which corresponds to a distance of 10 mm in the mould as the casting speed used is 0.5 m min^{-1}). The temperature profiles for this case are shown in Figure 5.5.17. It can be observed that the corner temperature is always higher than that at the mid-faces in the mould region. This is to be expected in view of the lower heat flux at the corner. The temperature at the middle of the long face just above the exit of the mould is lower than the temperature at the middle of the short face in view of higher heat extraction from that face. Reheating in the air gap near the bottom of the mould is clearly demonstrated. The solidification front for this case is shown in Figure 5.5.20. The calculated average shell thickness at the end of the mould coincides with that calculated on the basis of heat balance. The solidified thickness of the metal is greater on the long face, which is to be expected in view of the lower temperatures along that face. The calculations are continued below the mould with the temperatures at the exit of the mould as the initial temperatures. The resulting temperature profiles are also shown in Figure 5.5.17. Reheating of the strand in the spray zone below the mould is clearly demonstrated. The calculations are continued up to a distance of about 150 cm below the meniscus level. The solidification fronts corresponding to the distances of 39, 78 and 144 cm below the mould are also shown in Figure 5.5.20.

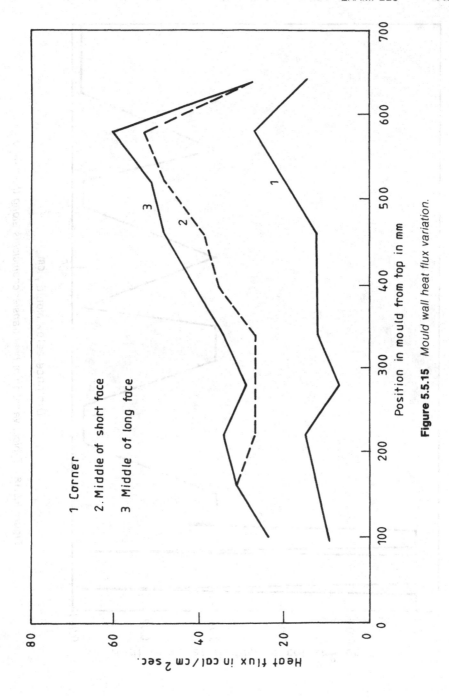

1 Corner

2. Middle of short face

3 Middle of long face

Heat flux in cal/cm² sec.

Position in mould from top in mm

Figure 5.5.15 *Mould wall heat flux variation.*

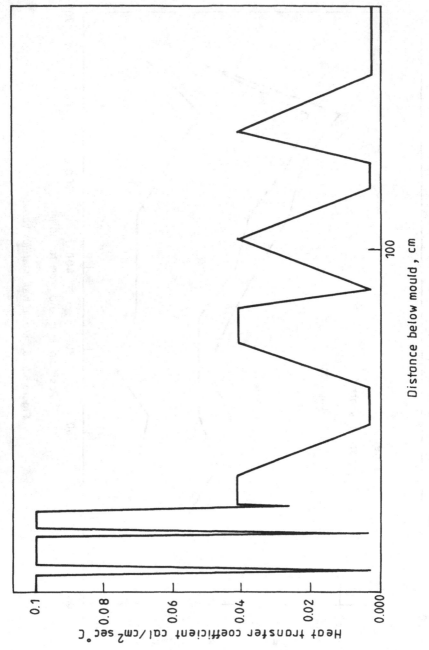

Figure 5.5.16 *Typical variation of heat transfer coefficients along the strand.*

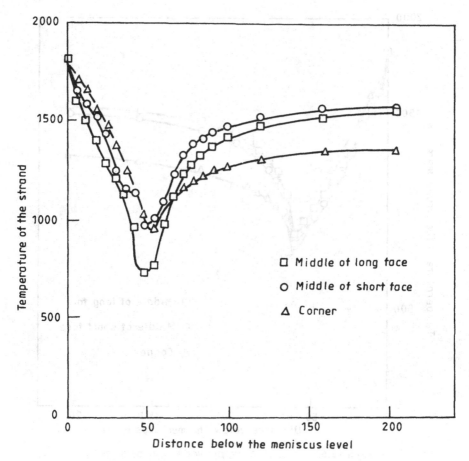

Figure 5.5.17 *Temperature profiles at various times.*

Case 2   In this case the calculations are repeated with each of the faces and corners assumed to have their respective individual average fluxes. The resulting temperature profiles are shown in Figure 5.5.18. It can be observed that in this case, there is no reheating in the strand in the air gap region of the mould. The temperatures predicted at the exit of the mould are lower than in the first case except for the long face. The temperatures on the short and long faces are nearly the same throughout, whereas the corner temperature is higher in the initial stages and is lower at the exit of the mould. The solidification front at the exit of the mould is shown in Figure 5.5.21. The calculated shell thickness is greater than that calculated from a heat balance. The solidification front on the long face coincides with that of Case 1, which is to be expected since the two temperatures are almost the same at the exit of the mould. However, the temperatures at the short face and corner are lower than in Case 1, which

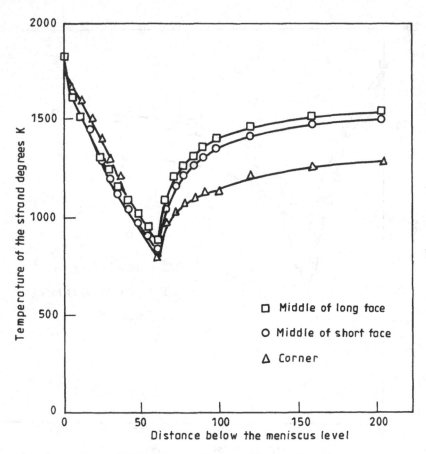

Figure 5.5.18   *Temperature profiles at various times.*

results in a greater shell thickness. Thus, the greater average shell thickness in this case is due to the lower temperatures at the short face and corner. The temperature profiles below the mould are also shown in Figure 5.5.18. It can be observed that the middle of the long face has a temperature profile that coincides with that of Case 1. The solidification fronts at distances of 39, 78 and 144 cm below the mould are also shown in Figure 5.5.21. The solidified thickness of the shell at a section 144 cm below the mould is greater than in Case 1.

Case 3   In this case the calculations are made with the average heat flux for the entire mould. The resulting temperature profiles on the strand are shown in Figure 5.5.19. It can be observed that the temperature at the corner is very low, a characteristic of using an average heat flux for the entire mould. In this case

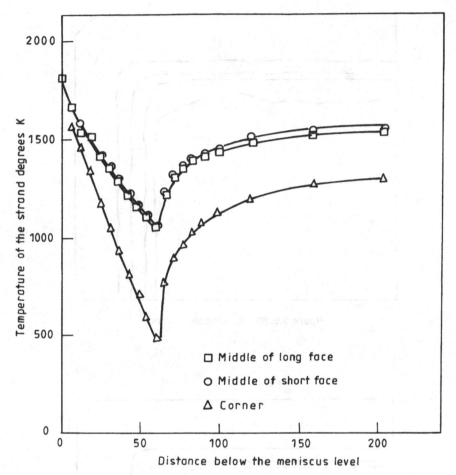

Figure 5.5.19 *Temperature profiles at various times.*

also, no reheat is predicted in the air gap of the mould region. Since an average heat flux is used for both the faces, the temperatures of both faces are almost the same. The temperature profiles of the strand below the mould indicate that, initially, the temperatures of the middle of both faces are higher than in Case 1, but gradually converge towards the values predicted in Case 1. The solidification fronts at distances of 0, 39 and 144 cm below the mould are shown in Figure 5.5.22. The shell thickness at the exit of the mould is only slightly higher than that calculated from a heat balance. It can be observed from the solidification front at the exit of the mould for this case that the shell thickness on the long and short faces are the same and both are equal to the shell thickness on the long face for Case 1. This is to be expected since the temperatures of these faces are the same. At the corner the shell thickness is greater than in

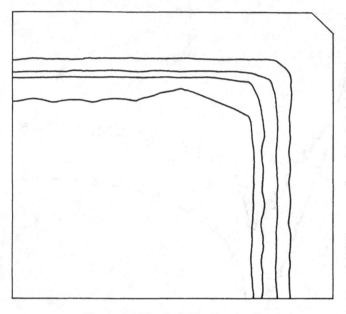

Figure 5.5.20   *Solidification fronts.*

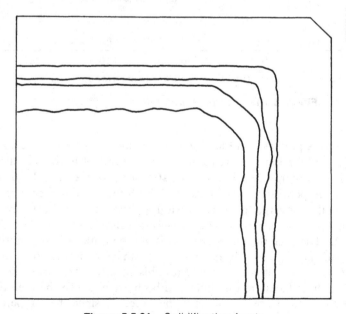

Figure 5.5.21   *Solidification fronts.*

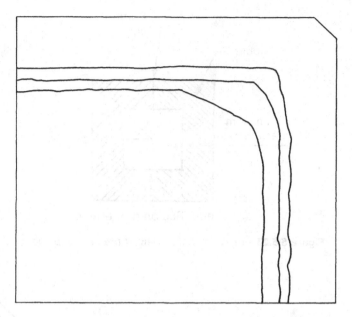

Figure 5.5.22 *Solidification fronts.*

Case 1 in view of a much lower corner temperature in the present case, which leads to a slightly greater average shell thickness.

From the analyses it can be concluded that the effect of the mould wall heat flux on the temperature of the strand after some distance below the mould is negligibly small which is in confirmation with the experimental observation.

5.5.4 *Aluminium casting*

Next, a square-shaped central section of a relatively long aluminium casting in a metallic mould is analysed as a two-dimensional heat transfer problem [22]. The experimental air gap data available in the literature are made use of [22]. The modified Newton–Raphson method [22] along with the Crank–Nicolson time-marching scheme, is used for the solution.

The central plane of a $70 \times 70\,mm^2$ aluminium casting of $280\,mm$ length (Figure 5.5.23) in a cast iron mould is discretised using linear triangular elements (Figure 5.5.24). Only one-eighth of the domain is discretised, taking advantage of symmetry of geometry and boundary conditions. A constant air gap of 1 mm is used to calculate the equivalent thermal conductivity. Combined convective and radiative boundary conditions are imposed on the outer surface of the mould. To ensure the accuracy of results a time step size of $0.05\,s$ is used. Figure

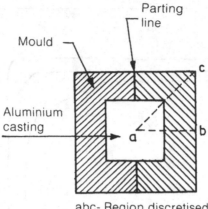

abc- Region discretised

Figure 5.5.23   *Cross-sectional view of the central plane.*

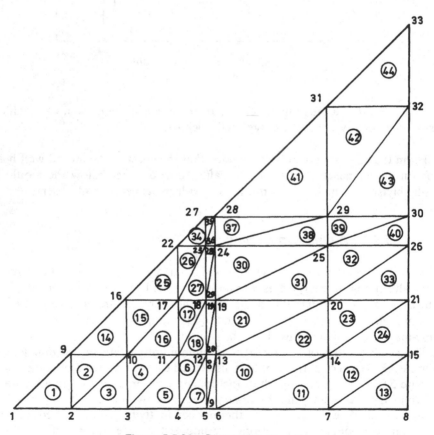

Figure 5.5.24   *Discretised domain.*

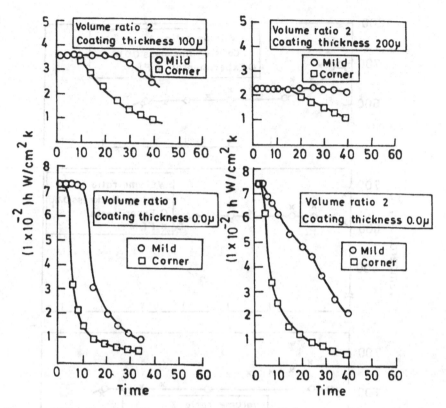

Figure 5.5.25 *Variation of heat transfer coefficient with time at mould–metal interface.*

5.5.25 shows the variations of heat transfer coefficient with time at the mould–metal interface [23].

The results computed in the present work are compared with the experimental results for different volume ratios and coating thicknesses for a few cases and found to be in good agreement (Figure 5.5.26). A decrease in volume ratio increases the solidification rate, whereas an increase in the coating thickness decelerates the solidification rate.

5.6 ADAPTIVE REMESHING IN SOLIDIFICATION PROBLEMS

The size and placement of the elements largely determine the accuracy with which the problem can be solved. Reducing the element sizes and thereby increasing the number of nodal points usually yields a more accurate solution, but at the cost of an increased CPU and memory requirement. The key to the

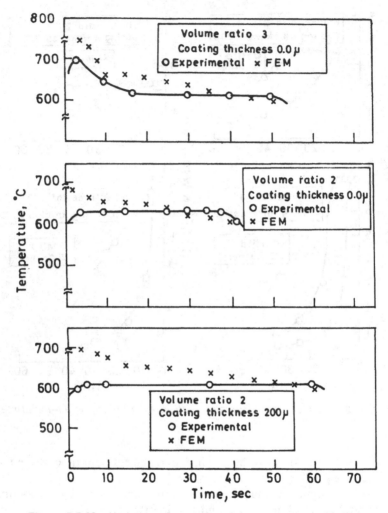

Figure 5.5.26 *Variation of central nodal temperatures with time.*

efficient and economic solution of problems is not merely the number of nodal points and elements but also their placement. An accurate numerical analysis will invariably require judgement on behalf of the user as to the placement of mesh density. Regions with large gradients (e.g. a discontinuity, stress concentration or region of high heat flux) will need a high mesh density, with quiescent regions requiring a comparatively coarser mesh. In many real engineering situations, attempts will be made either to obtain the most accurate solution possible with an upper limit on problem size, or to capture a numerically awkward feature without using an excessive number of elements. In addition, the adaptive remeshing technique will be helpful to an industrial user who need not be concerned with optimising a good starting mesh which requires an idea of the nature of the solution of the problem. This removes from the user the drudgery of discretisation.

The premise of adaptive procedures is that by making use of the mathematics of error analysis, a finite element program can inform the user which regions need refining and automatically adapt the mesh to suit the problem. The error is calculated in each element and is compared with a predefined limit. Any element with an error above, and in some cases below, the limit, must be adjusted in size. The process is repeated, if necessary, with the ultimate aim that every element contains the same predefined, allowable error, thus yielding an optimal mesh.

A simplified error estimate is outlined in a very useful paper by Zienkiewicz and Zhu [24] which does not require the calculation of inter-element traction jumps, and which was shown to give a satisfactory approximation to that of the error estimate reported by Kelly *et al.* [25]. The remeshing approach involves completely regenerating the mesh, either in regions of high error only or over the entire domain. Triangles are the optimal shapes because of the high degree of flexibility, and the advancing front method [26] has been a popular and successful method of generating such elements [26, 27]. Huang and Lewis [28] were the first to give an adaptive analysis for heat flow problems using the error estimation technique. Lewis *et al.* [29] extended the work reported in [28] to non-linear transient heat conduction problems. Here the error is defined in terms of temperature gradients rather than temperatures, and for a two-dimensional element it is written as

$$e_e = \left[\int_{\Omega_e} \left[k_x \left(\frac{\partial \bar{T}}{\partial x} - \frac{\partial \tilde{T}}{\partial x} \right)^2 + k_y \left(\frac{\partial \bar{T}}{\partial y} - \frac{\partial \tilde{T}}{\partial y} \right)^2 \right] dx\, dy \right]^2 \qquad (5.6.1)$$

where $\partial \bar{T}/\partial x$ represents a smoothed gradient and $\partial \tilde{T}/\partial x$ represents a non-smooth gradient. Now this error is compared for all elements to the maximum permissible error in an element as calculated above and used to modify the mesh for a second analysis. A variable, ξ, is defined for an element as

$$\xi_e = \frac{\| e \|_e}{\| \bar{e} \|_e} \qquad (5.6.2)$$

where

$$\| e \| \leqslant \bar{\eta} \left(\frac{\| q \|^2}{m} \right) \qquad (5.6.3)$$

$m = $ the number of elements and

$$\| q \|^2 \simeq \| \tilde{q} \|^2 = \sum_{e=1}^{m} \int_{\Omega_e} \left[k_x \left(\frac{\partial \tilde{T}}{\partial x} \right)^2 + k_y \left(\frac{\partial \tilde{T}}{\partial y} \right)^2 \right] dx\, dy \qquad (5.6.4)$$

If $\xi_e > 1$, the size of the element e must be reduced, and vice versa. If the existing element size is h_e, the new element size, $\bar{h}_e$, is given by

$$\bar{h}_e = \frac{h_e}{\xi_e^{1/p}} \tag{5.6.5}$$

where p is generally taken as the order of the shape functions. When the information regarding all element sizes is available, a new mesh is generated on which another analysis may be carried out. This procedure is continued until the calculated error falls below the specified maximum.

The application of adaptive remeshing to a practical problem of solidification in casting is considered next. Figure 5.6.1(a) shows the details of a typical casting problem involving the solidification of liquid metal poured into a sand mould. The initial finite element mesh for this problem is shown in Figure 5.6.1(b). A thin metal fin protrudes from the main body of the casting into the sand mould. The initial temperature of the metal is $700\,°C$ and that of the sand is $20\,°C$. The latent heat of solidification for the metal is accounted for by the enthalpy method. This configuration, though not very complicated, does not immediately suggest the type of mesh that must be used for a reasonable solution. For this particular problem the choice of a good mesh is further complicated by the fact that a solidification front exists which moves as the transient solution evolves and the latent heat is released. An adaptive remeshing procedure is well suited to such problems. Since this is a transient problem the adaptive procedure is varied slightly. The approach preferred is to check the error after a fixed number of time steps and to modify the mesh accordingly. The additional work involved

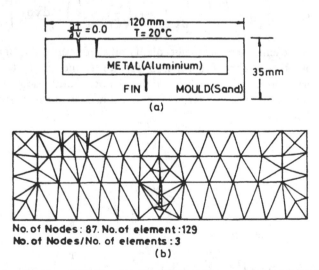

No. of Nodes: 87. No. of element: 129
No. of Nodes/No. of elements: 3
(b)

Figure 5.6.1 *Problem definition for the solidification problem and the initial mesh.*

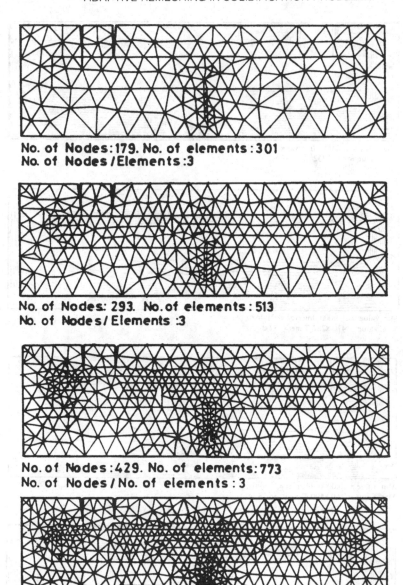

No. of Nodes:179. No. of elements:301
No. of Nodes/Elements:3

No. of Nodes: 293. No. of elements:513
No. of Nodes/Elements :3

No. of Nodes:429. No. of elements:773
No. of Nodes/No. of elements:3

No. of Nodes:597. No. of elements:1094
No. of Nodes/Elements:3

Figure 5.6.2 *Meshes for the transient heat conduction problem for the* 10th, 11th, 20th, 30th *and* 40th *time steps.*

Min. value = 20.0000 Interval = 3? 9749
Max. value = 672.5732 Time = 5.16

Min. value = 20.0000 Interval = 21 4680
Max. value = 470.8287 Time = 37.07

Min. value = 20.0000 Interval = 7 7361
Max. value = 169.8590 Time = 55.16

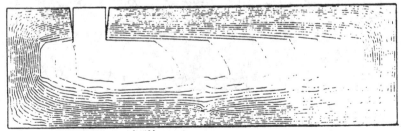

Min. value = 20.0000 Interval = 2 4853
Max. value = 72.1905 Time = 135.16

Figure 5.6.3   *Temperature contours for the transient heat conduction problems for the 10th, 11th, 20th, 30th and 40th time steps.*

here is the need to interpolate the nodal information from the previous mesh to the new mesh in order to continue the time integration process. For the present example the mesh is regenerated after every ten steps. Figure 5.6.2 shows the refined meshes obtained from the use of the adaptive procedure, at the 10th, 20th, 30th and 40th time steps. The corresponding temperature contours at the same time levels are shown in Figure 5.6.3. The norm error, $\bar{\eta}$, to be achieved is specified as 40%. The norm errors after every ten steps before refining the mesh are found to be 72.9%, 41.0%, 32.1% and 25.1%, which correspond to temperature errors of approximately 53.16%, 16.8%, 10.3% and 6.3%, respectively. The mesh refinement is to be concentrated at and near the protruding fin and mould–metal boundaries, where a concentration of temperature gradients exists owing to the significant difference in thermal properties of the sand and metal.

5.7 OTHER APPLICATIONS

The finite element method has also been extended to other physical phenomena involving a change of phase. One example is the analysis of a cabinet specially designed for the protection of floppy disks, valuable documents, etc. against fire hazard [30]. Another involves the solidification of molten metal during a process known as 'squeeze forming' which is extensively used for producing high integrity components such as wheels for military vehicles [31].

REFERENCES

[1] Crank J., *Free and Moving Boundary Problems* Clarendon Press, Oxford (1984).
[2] Dalhuijsen A. J. and Segal A., Comparison of finite element techniques for solidification problems, *Int. J. Num. Meth. Engg*, **23**, 1807–1829 (1986).
[3] Salcudean M. and Abdullah Z., On the numerical modelling of heat transfer during solidification processes, *Int. J. Num. Meth. Engg*, **25**, 445–473 (1988).
[4] Voller V. R., Swaminathan C. R. and Thomas B. G., Fixed grid techniques for phase change problems, a review, *Int. J. Num. Meth. Engg*, **30**, 875–898, Special Thermal Issue (1990).
[5] Del Guidice S., Comini G. and Lewis R. W., Finite element simulation of freezing processes in solids, *Int. J. Num. Analy. Meth. Goemechanics*, **2**, 223–235 (1978).
[6] Lemmons, E.C., Multidimensional integral phase change approximations for finite element conduction codes, In: *Numerical Methods in Heat Transfer*, Eds. R. W. Lewis, K. Morgan and O. C. Zienkiewicz John Wiley & Sons, Chichester (1981).
[7] Morgan K., Lewis R. W. and Zienkiewicz O. C., An improved algorithm for heat conduction problems with phase change, *Int. J. Num. Methods Engg*, **12**, 1191–1195 (1978).
[8] Lewis R. W. and Roberts P. M., Finite element simulation of solidification problems, *Appl. Scientific Res.*, **44**, 61–92 (1987).
[9] Ralph W. D. and Bathe K. J., An efficient algorithm for analysis of nonlinear heat transfer with phase changes, *Int. J. Num. Meth. Engg*, **18**, 119–134 (1982).

[10] Usmani A., *Finite Element Modelling of Convective–Diffusive Heat Transfer and Phase Transformation with Reference to Casting Simulation*, Ph.D. Dissertation, University College of Swansea, U.K. (1991).

[11] Carslaw H. S. and Jaeger J. C., *Conduction of Heat in Solids* Clarendon Press, Oxford (1959).

[12] Massey I. D. and Sheridan A. T., Theoretical predictions of earliest rolling times and solidification times of ingots, *J. Iron Steel* Inst., **209**, no. 5, 391 (1971).

[13] Sevrin R., *Mathematical Study of the Thermal Evolution of Steel Ingots, Mathematical Models in Metallurgical Process Development* Iron and Steel Institute, London, pp. 147–157 (1970).

[14] Saliman J. R. and Fakhroo E. A., Finite elements solutions of Heat transmission in steel ingots, *J. Mech. Engg, Sci.*, **14**, no. 1, 19–24 (1972).

[15] Comini G., Del Guidice S., Lewis R. W. and Zienkiewicz O. C., Finite element solution of nonlinear heat conduction problems with special reference to phase change, *Int. J. Num. Meth. Engg*, **8**, 613–624 (1974).

[16] Lees M., A linear three-level difference scheme for quasilinear parabolic equations *math. Comp.*, **20**, 515–522 (1966).

[17] Weingart H., *Untersuchungen uber den Temperatur-und Erstarrung sablauf shewerer Blocke und Brammen' Investigations into the cooling and solidification of Heavy Ingots and Slabs* Dissertation, Clausthal, West Germany (1968).

[18] den Hartog H. W. Rabenberg J. M. and Willemse J., Application of a mathematical model on the study of the ingot solidification process *Iron Making Steel Making*, **2**, 134–144 (1975).

[19] Lewis R. W., Seetharamu K. N. and Morgan K., Application of the finite element method in the study of ingot castings *Solidification Technology in the Foundry Cast House*, Proc. Int. Conf., Univ. of Warwick, Coventry Metals society of London, pp. 40–43 (1980).

[20] Oeters, P., Ruthger, K. and Selenz, K. J., Heat transfer in ingot pouring *Information Symposium on Casting and Solidification of Steel, Luxembourg* vol 1, pp. 126–167 (1977).

[21] Morgan K., Lewis R. W. and Seetharamu K, N., Effect of mould wall heat flux variation on solidification in continuous casting of steel *Int. Conf. Nonlinear Problems* Swansea (1980).

[22] Naik H. B. and Dave A. K., Solidification of castings in metallic moulds *Comm. Appl. Num. Meth.*, **5**, 467–472 (1989).

[23] Isaac J., *Experimental and Theoretical Studies on Solidification of Castings in Metallic Moulds* Ph.D. Thesis, I.I.T. Bombay (1983).

[24] Zienkiewicz O. C. and Zhu J. Z., A simple error estimator and adaptive procedure for practical engineering analysis *Int. J. Num. Meth. Engg*, **24**, 337–352 (1987).

[25] Kelly D., Gago J. D. S., Zienkiewicz O. C. and Babuska I., A posterior error analysis and adaptive processes in the finite element method, Part 1—Error analysis *Int. J. Num. Meth. Engg*, **19**, 1596–1619 (1983).

[26] Peraire J., Vahdati M., Morgan K. and Zienkiewicz O. C., Adaptive remeshing for compressible flow computations *J. Comp. Phys.*, **72**, 449–466 (1987).

[27] Cross J. T., *A Two-Dimensional Triangular Mesh Generator Using the Advancing Front Method* Technical Report CR/662/91, Dept of Civil Engg, University College of Swansea, U.K. (1990).

[28] Huang H. C. and Lewis R. W., Adaptive analysis for heat flow problems using error estimation techniques *Sixth Int. Conf. for Numerical Methods in Thermal Problems*, Swansea, UK, Pineridge Press (1989).

[29] Lewis R. W., Huang H. C., Usmani A. S. and Cross J. T., Finite element analysis of heat transfer and flow problems using adaptive remeshing including application to solidification problems *Int. J. Num. Meth. Engg*, **32**, 767–781 (1991).

[30] Lewis R. W., Huang H. C. and Welch T. N., A finite element analysis of fire-resisting cabinets using an adaptive remeshing technique *Appl. Math. Modelling*, **15**, 274–279 (1991)

[31] Gethin D. T., Lewis R. W. and Tadayon M. R., A finite element approach for modelling metal flow and pressurised solidification in the squeeze casting process *Int. J. Num. Meth. Engg*, **35**, 939–950 (1992).

[32] Morgan K., Lewis R. W. and Seetharamu, K. N. Modelling heat flow and thermal stress in ingot casting, *Simulation*, **36**, 55–63 (1981).

6 Convective Heat Transfer

6.1 INTRODUCTION

For the successful design of many engineering systems, the ability to predict the rate of heat transfer between a fluid medium and a solid boundary is essential. Nuclear reactors, space power plants, super computers, superconducting magnets, re-entry vehicles, combustion chambers and gas turbine blade cooling are some typical examples where a knowledge of the rate of heat transfer between a fluid and solid is extremely important. The solutions of these types of problems usually require a mathematical description that is too complex to allow an analytical determination of the fluid and thermal fields. Considerable effort has been directed towards the development of approximate solution methods by using various numerical methods. During the 1970s and 1980s research in the use of finite element methods for fluid mechanics allowed the extension of this method to problems in convective heat transfer. Convection problems are normally divided into two broad categories, namely forced convection and free or natural convection. In forced convection, the fluid motion is due to the applied pressure or viscous forces on the fluid boundary. When the fluid motion is produced by buoyancy forces, which are usually induced by temperature gradients but may also be due to the concentration gradients, the process is termed free or natural convection. In this chapter the fluid is assumed to be incompressible, i.e. the variation of density with temperature is neglected (except for the case of the buoyancy term in natural convection). In the following sections the basic equations describing the fluid flow and thermal energy transport in the continuum are presented. The finite element approximations to these equations will then be described. Some examples are given of the convection problems that are encountered in practice.

6.2 BASIC EQUATIONS

In order to simplify the problem, a number of assumptions and restrictions are made. The fluid is assumed to be Newtonian and incompressible. The flow is considered to be laminar. The material properties are allowed to vary with temperature. The effects of radiative transport and viscous dissipation (bacause of low velocity flows) are neglected. The starting point for the analysis are the laws of conservation of mass, momentum and energy [1].

Mass:

$$\frac{\partial u}{\partial x} + \frac{\partial v}{\partial y} + \frac{\partial w}{\partial z} = 0 \qquad (6.2.1)$$

Momentum

$$\rho \frac{Du}{Dt} = -\rho \beta g_x (T - T_\infty) + \frac{\partial}{\partial x}(\sigma_x - p) + \frac{\partial}{\partial y}\tau_{xy} \qquad (6.2.2)$$

$$\rho \frac{Dv}{Dt} = -\rho \beta g_y (T - T_\infty) + \frac{\partial}{\partial x}\tau_{xy} + \frac{\partial}{\partial y}(\sigma_y - p) \qquad (6.2.3)$$

Energy:

$$\rho c \frac{DT}{Dt} = \frac{\partial}{\partial x}\left(k_x \frac{\partial T}{\partial x}\right) + \frac{\partial}{\partial y}\left(k_y \frac{\partial T}{\partial y}\right) + G \qquad (6.2.4)$$

where

$$\frac{D}{Dt} = \frac{\partial}{\partial t} + u\frac{\partial}{\partial x} + v\frac{\partial}{\partial y}$$

u = velocity in the x direction
v = velocity in the y direction
ρ = density of the fluid
p = pressure
T = temperature
c = heat capacity
k = thermal conductivity
μ = viscosity
β = coefficient of volume expansion
G = volumetric heat source
g_x, g_y = acceleration components
σ_x, σ_y = normal stresses
τ_{xy} = shear stress

The number of unknowns in the above system of equations is four: u, v, P and T. There are four equations to solve.

The boundary conditions are

$$u = g(x, y) \qquad v = j(x, y) \quad \text{on } S_1 \qquad (6.2.5a)$$

Surface tractions

$$\left.\begin{array}{l} \bar{\sigma}_x = (\sigma_x - p)\eta_x + \tau_{xy}\eta_y \\ \bar{\sigma}_y = \tau_{xy}\eta_x + (\sigma_y - p)\eta_y \end{array}\right\} \text{ on } S_2 \qquad \begin{array}{l} (6.2.5b) \\ (6.2.5c) \end{array}$$

$$T = f(x, y) \quad \text{on } S_3 \tag{6.2.5d}$$

$$-k_x \frac{\partial T}{\partial x} \eta_x - k_y \frac{\partial T}{\partial y} \eta_y + q + h(T - T_\infty) = 0 \quad \text{on } S_5 \tag{6.2.5e}$$

By introducing the stress components

$$\sigma_x = 2\mu \frac{\partial u}{\partial x}$$

$$\sigma_y = 2\mu \frac{\partial v}{\partial y} \tag{6.2.6}$$

$$\tau_{xy} = \mu \left[\frac{\partial u}{\partial y} + \frac{\partial v}{\partial x} \right]$$

into equations (6.2.2) and (6.2.3), we get, for the momentum equations:

$$\rho \frac{Du}{Dt} = -\frac{\partial p}{\partial x} + \mu \left(\frac{\partial^2 u}{\partial x^2} + \frac{\partial^2 u}{\partial y^2} \right) - g_x \beta \rho (T - T_\infty) \tag{6.2.7}$$

$$\rho \frac{Du}{Dt} = -\frac{\partial p}{\partial y} + \mu \left(\frac{\partial v}{\partial x^2} + \frac{\partial v}{\partial y^2} \right) - g_y \beta \rho (T - T_\infty) \tag{6.2.8}$$

6.3 GALERKIN METHODS FOR STEADY CONVECTION–DIFFUSION PROBLEMS

We consider a one-dimensional steady convection-dominated flow given by (where u is known) [2]

$$u \frac{dT}{dx} = \alpha \frac{d^2 T}{dx^2} \tag{6.3.1}$$

This equation corresponds to the flow of a liquid metal in which the thermal boundary layer is far greater than the hydrodynamic boundary layer and hence the velocity u in the thermal boundary layer is constant. α is the thermal diffusivity of the liquid metal (where $\alpha = k/\rho c$). The discretised mesh is of uniform length and two-noded linear elements are used:

$$T = T_1 N_1 + N_2 T_2 \tag{6.3.2}$$

$$\frac{dT}{dx} = -\frac{T_i}{l} + \frac{T_j}{l}$$

Applying the standard Galerkin method to equation (6.3.1), we have

$$\int_l \left(\frac{u \, dT}{\alpha \, dx} - \frac{d^2 T}{dx^2} \right) N_i \, dl = 0 \quad i = 1, 2 \tag{6.3.3}$$

$$-\frac{dT}{dx}\bigg|_0 + \frac{dT}{dx}\bigg|_l + \frac{u}{2\alpha}\begin{bmatrix} -1 & 1 \\ -1 & 1 \end{bmatrix}\begin{Bmatrix} T_i \\ T_j \end{Bmatrix} + \frac{1}{l}\begin{bmatrix} 1 & -1 \\ -1 & 1 \end{bmatrix}\begin{Bmatrix} T_i \\ T_j \end{Bmatrix} = 0 \quad (6.3.4)$$

This equation has four unknowns: T_i, T_j, $dT/dx|_0$ and $dT/dx|_l$.

When we combine the equations for m elements for a given problem, the two boundary conditions of flux density at the two ends are substituted. Thus there are $(m+1)$ equations and $(m+1)$ unknowns.

Assembly of equation (6.3.4) to two elements of equal size l (ignoring the effect of dT/dx at the ends):

$$\frac{u}{2\alpha}\begin{bmatrix} -1 & 1 & 0 \\ -1 & 0 & 1 \\ 0 & -1 & 1 \end{bmatrix}\begin{Bmatrix} T_1 \\ T_2 \\ T_3 \end{Bmatrix} + \frac{1}{l}\begin{bmatrix} 1 & -1 & 0 \\ -1 & 2 & -1 \\ 0 & -1 & 1 \end{bmatrix}\begin{Bmatrix} T_1 \\ T_2 \\ T_3 \end{Bmatrix} = 0 \quad (6.3.5)$$

For an interior node, we get

$$(1 + P_e)T_{i-1} - 2T_i + (1 - P_e)T_{i+1} = 0 \quad (6.3.6)$$

where

$$P_e = \frac{ul}{2\alpha} = \text{element Peclet number} \quad (6.3.7)$$

and where T_{i-1}, T_i and T_{i+1} are three consecutive nodes. If a central difference finite difference scheme is used, we get the same equation, namely equation (6.3.6).

Rigorous mathematical analysis shows that equation (6.3.6) can be written as

$$\left[-u\frac{dT}{dx} + \alpha\frac{d^2T}{dx^2} \right]_i + \frac{\alpha}{2P_e}\left[\frac{1}{P_e}(\cosh 2P_e - 1)\sinh 2P_e \right]\left(\frac{d^2T}{dx^2} \right)_i = 0 \quad (6.3.8)$$

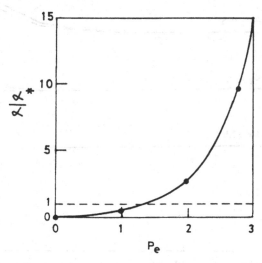

Figure 6.3.1 α^*/α as a function of element Peclet number, P_e. (Reproduced with permission from Donea J., Recent advances in computational methods for steady and transient transport problems. Nucl. Engg. Des., **80**, no. 2, 141–162 (1984), © Elsevier Sequoia Lausanne.)

It can be observed that in equation (6.3.8) the first term is the original differential equation (6.3.1) and the second term is the truncation error in the numerical scheme which has the character of diffusion. Equation (6.3.8) can be written as

$$U\frac{dT}{dx} = (\alpha - \alpha^*)\frac{d^2T}{dx^2} \tag{6.3.9}$$

where, according to equation (6.3.8),

$$\alpha^* = \frac{-\alpha}{2P_e}\left[\frac{1}{P_e}(\cosh P_e - 1) - \sinh 2P_e\right] \tag{6.3.10}$$

Thus, we observe that the discrete equation (6.3.6) obtained by the standard Galerkin method applied to equation (6.3.1) solves a modified under-diffused

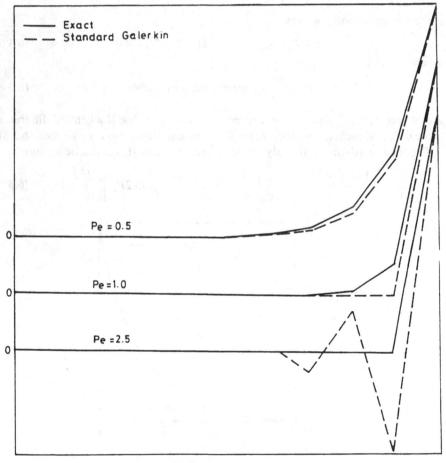

Figure 6.3.2 *Effect of element Peclet number on the solution. (Reproduced with permission from Donea J., Recent advances in computational methods for steady and transient transport problems. Nucl. Engg. Des., **80**, no. 2, 141–162 (1984), © Elsevier Sequoia Lausanne.)*

equation (6.3.9). A plot of α^*/α versus P_e (shown in Figure 6.3.1) indicates that $\alpha^* > \alpha$ for $P_e > 1$ and thus we are solving an equation with negative diffusion. Hence problems may be expected in the numerical results for $P_e > 1$, which are shown in Figure 6.3.2 for a specified temperature at the outlet [3]. The remedy for this problem is to refine the mesh in such a way that the element Peclet number is less than 1 [4]. However, the fact remains that the standard Galerkin finite element solutions are *always under diffused* for all element Peclet numbers. This basic deficiency shows that the standard Galerkin method is not the optimal method for solving the convection-dominated flows.

6.4 UPWIND FINITE ELEMENTS IN ONE DIMENSION

The standard Galerkin finite element method for convection–diffusion problems always gives under diffused solutions; the remedy lies in adding artificial diffusion to the existing diffusion. In the context of the finite difference, the convective term is written in terms of upwind differencing, namely shifting the first derivative

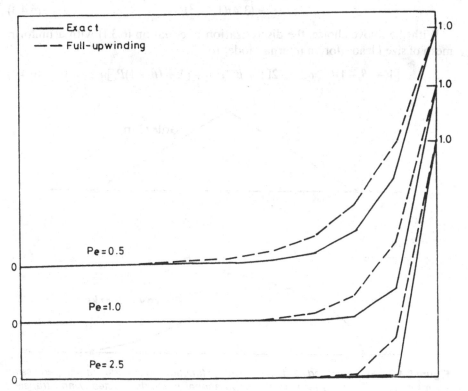

Figure 6.4.1 *Effect of upwinding on the solution. (Reproduced with permission from Donea J., Recent advances in computational methods for steady and transient transport problems. Nucl. Engg. Des., **80**, no. 2, 141–162 (1984), © Elsevier Sequoia Lausanne.)*

term in the upstream direction. The upwind method gives solutions that are over diffused, as shown in Figure 6.4.1. Thus a suitable combination of the above schemes should provide an optimal solution.

Several methods have been developed to achieve the upwind effect in finite element formulations of steady-convection–diffusion problems.

6.4.1   Petrov–Galerkin formulation using special weighting functions

Weighting functions W_i are modified to weight the element upwind of node i more heavily than the downwind element. One such modified weighting function is shown in Figure 6.4.2:

$$W_1(\xi) = N_1(\xi) - \beta F(\xi) \tag{6.4.1}$$
$$W_2(\xi) = N_2(\xi) + \beta F(\xi) \tag{6.4.2}$$

where β is a free parameter and

$$F = (3/4)(1 - \xi^2) \tag{6.4.3}$$

With the above choice, the discretization of equation (6.3.1) with a uniform mesh of size l leads (for an internal node) to

$$[1 + (\beta - 1)P_e]u_{i+1} - 2[1 + \beta P_e]u_i + [1 + (\beta + 1)P_e]u_{i-1} = 0 \tag{6.4.4}$$

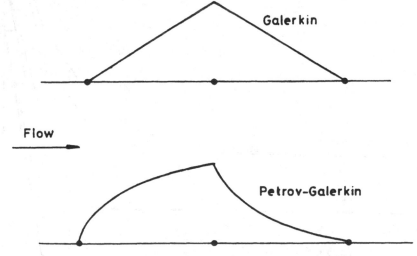

Figure 6.4.2 *Standard and modified weighting functions. (Reproduced with permission from Donea J., Recent advances in computational methods for steady and transient transport problems. Nucl. Engg. Des., **80**, no. 2, 141–162 (1984), © Elsevier Sequoia Lausanne.)*

Mathematical analysis shows that this is equivalent to

$$\left[-a\frac{du}{dx} + \alpha\frac{d^2u}{dx^2} \right]_i + \frac{\alpha}{2P_e}\left[\left(\frac{1}{P_e} + \beta\right)(\cosh 2P_e - 1) - \sinh 2P_e \right]\left(\frac{d^2u}{dx^2}\right)_i = 0$$

(6.4.5)

The second term in equation (6.4.5) is the exact truncation error of the discrete equation (6.4.4) and thus error can be reduced to zero by selecting the parameter β according to

$$\left(\frac{1}{P_e} + \beta\right)(\cosh 2P_e - 1) = \sinh P_e$$

which leads to [5]

$$\beta = \coth P_e - \frac{1}{P_e}$$

(6.4.6)

which gives the exact solutions for all values of the element Peclet number.

A new Petrov–Galerkin method is presented by De Sampio [6] which is derived from the concept of using a modified function to make the differential operator self-adjoint. The value of the optimal upwind parameter β is derived from the process of approximating the modifying function.

The new weighting functions [6] are given by

$$W_1 = N_1(1 - \beta)$$ (6.4.7)
$$W_2 = N_2(1 + \beta)$$ (6.4.8)

where $(1 - \beta)$ and $(1 + \beta)$ are modifying functions. W_1 and W_2 make use of the piece-wise constant discrete modifying functions $(1 - \beta)$ and $(1 + \beta)$, respectively.

A one-dimensional steady-state problem with constant source term f and homogeneous boundary conditions at $x = 0$ and $x = l$ is considered. The analytical solution is given by

$$u = \frac{fl}{\rho C_p a}\left\{ \lambda\frac{x}{l} - \frac{\exp\left[P_e\left(\frac{x}{l} - 1\right)\right] - \exp(-P_e)}{1 - \exp(-P_e)} \right\}$$

(6.4.9)

with $P_e = \rho C_p al/k$.

The present solution gives nodally exact solutions on the uniform mesh. Figure 6.4.3 shows the solution on the non-uniform mesh for $P_e = 20$ and 100 along with the analytical solution indicated by the continuous line. The element Peclet number, P_e, varies from 1 to 3 in the first case and from 5 to 15 in the second case. The analytical solution represents a boundary-layer type at $x = L$, which becomes thinner as P_e increase. The mesh requires a refinement at $x = L$. The results obtained are nodally exact even for the non-uniform meshes.

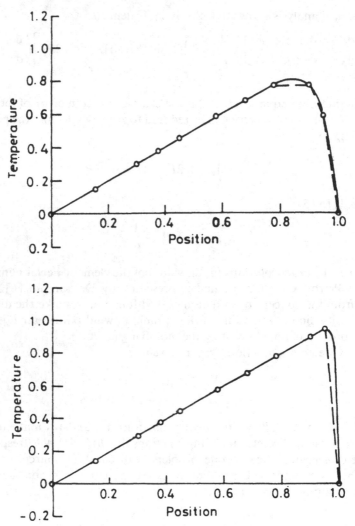

Figure 6.4.3 *Solution on non-uniform mesh $P_e = 20$ (top) and $P_e = 100$ (bottom). (Reproduced with permission from Donea J., Recent advances in computational methods for steady and transient transport problems. Nucl. Engg. Des., **80**, no. 2, 141–162 (1984), © Elsevier Sequoia Lausanne.)*

6.4.2 Upwind schemes based on a modified quadrature rule

The Petrov–Galerkin algorithm given in equations (6.4.1) and (6.4.2) involves quadratic weighting functions and hence requires higher-order quadrature rules, which are expensive. Hughes [7] has developed an optimal upwind finite element scheme using piece-wise linear shape and test functions in conjunction with a special quadrature, thereby making the algorithm simpler to implement and computationally more efficient. A single quadrature point, β, positioned within the element according to

$$\beta = \coth P_e - \frac{1}{P_e}$$

is used in evaluating the integral for the convective term.

6.4.3   Upwind schemes based on the artificial diffusion interpretation

An artificial diffusion $\bar{\alpha}$, given by

$$\bar{\alpha} = \beta \frac{al}{2} \tag{6.4.10}$$

is added to the physical diffusion and a conventional Galerkin finite element discretisation is employed. Hence, we get, using the Galerkin method

$$\int_0^l \left[wa\frac{du}{dx} + \frac{dw}{dx}(\alpha + \bar{\alpha})\frac{du}{dx} \right] dx = 0 \tag{6.4.11}$$

In view of $\bar{\alpha} = \beta(al/2)$, we get

$$\int_0^l \left[\left(w + \frac{\beta l}{2}\frac{dw}{dx} \right) a\frac{du}{dx} + \frac{dw}{dx}\alpha\frac{du}{dx} \right] dx = 0 \tag{6.4.12}$$

Equation (6.4.12) indicates that the added diffusion method can also be regarded as a Petrov–Galerkin method in which a modified weighting function given by

$$\bar{w} = w + \frac{\beta l}{2}\frac{dw}{dx} \tag{6.4.13}$$

is applied to the convective term. The weighting function given by equation (6.4.13) is discontinuous across element boundaries.

6.4.4   Upwind method extended to multi dimensions

The upwind finite element methods described in the earlier solutions provided a significant improvement over the conventional Galerkin method for the simple one-dimensional, linear convection–diffusion problem, equation (6.3.1). Problems have been identified while extending the upwind methods to the multidimensional problem and methods have been devised to overcome these difficulties to varying degrees [8, 9].

6.5 *TAYLOR–GALERKIN METHOD FOR TRANSIENT CONVECTION–DIFFUSION PROBLEMS*

The upwind method has essentially been derived based on the steady-state convection–diffusion equation and is not strictly applicable to the transient equations. In addition, the upwind method still requires empirical factors to control the amount of artificial diffusion.

A better alternative appeared to be the Taylor–Galerkin methods forwarded by Donea [10] who developed these methods for different time-marching schemes. Donea's work was motivated by the work of Morton and Parrot [11] who showed that to each particular time stepping method, corresponds a different optimal form of the Petrov–Galerkin weighting functions. This led him to discretise the pure convection equation in time first, with an improved difference approximation of the time derivative term by including higher order Taylor series terms, and then using the conventional Bubnov–Galerkin Finite Element Method (FEM) for spatial discretisation, thus leading to the 'Taylor–Galerkin' method. This method is the finite element counterpart of the Lax–Wendroff schemes used in finite differences [12]. Löhner *et al.* [13] used an Eulerian–Lagrangian approach to arrive at the same system of discretised equations obtained by Donea using the Taylor–Galerkin techniques, and justified the use of Galerkin's Finite Element Method (GFEM) for spatial discretisation by writing the convection equation in Lagrangian coordinates and obtaining an adjoint equation for which the conventional Bubnow–Galerkin FEM is optimal. This confirmed the characteristic-based nature of the Taylor–Galerkin technique.

Gresho *et al.* [14] derived the same scheme from the agrument of 'negative diffusion' that is generated by forward Euler time stepping and called their correction balancing tensor diffusivity. Donea *et al.* [15] have extended the Taylor–Galerkin method to the advection–diffusion problems.

Consider the convection–diffusion equation:

$$\rho c \left(\frac{\partial T}{\partial t} + u \frac{\partial T}{\partial x} \right) = k \frac{\partial^2 T}{\partial x^2} \tag{6.5.1}$$

Expand T using forward Taylor series:

$$T_{n+1} = T_n + \frac{\partial T_n}{\partial t} \Delta t + (1/2!) \frac{\partial^2 T_n}{\partial t^2} (\Delta t)^2 + O(\Delta t)^3$$

Therefore

$$\frac{T_{n+1} - T_n}{\Delta t} = \frac{\partial T_n}{\partial t} + \frac{\Delta t}{2} \frac{\partial^2 T_n}{\partial t^2} + O(\Delta t)^2 \tag{6.5.2}$$

From (6.5.1), we get

$$\rho c \frac{\partial T_n}{\partial t} = - \rho c u \frac{\partial T_n}{\partial x} + k \frac{\partial^2 T_n}{\partial x^2} \tag{6.5.3}$$

$$\rho c \frac{\partial^2 T_n}{\partial t^2} = -\rho c u \frac{\partial}{\partial x}\left(\frac{\partial T_n}{\partial t}\right) + k \frac{\partial^2}{\partial x^2}\left(\frac{\partial T_n}{\partial t}\right) \tag{6.5.4}$$

Substituting (6.5.3) and (6.5.4) in (6.5.2) yields the semi-discrete equation

$$\frac{T_{n+1} - T_n}{\Delta t} = \left(-u\frac{\partial T_n}{\partial x} + \frac{k}{\rho c}\frac{\partial^2 T_n}{\partial x^2}\right) + \frac{\Delta t}{2}\left[-u\frac{\partial}{\partial x}\left(\frac{\partial T_n}{\partial t}\right)\right.$$

$$\left. + \frac{k}{\rho c}\frac{\partial^2}{\partial x^2}\left(\frac{\partial T_n}{\partial t}\right)\right] + O(\Delta t)^2 \tag{6.5.5}$$

$$= \left(-u\frac{\partial T_n}{\partial x} + \frac{k}{\rho c}\frac{\partial^2 T_n}{\partial x^2}\right) + \frac{\Delta t}{2}\left[-u\frac{\partial}{\partial x}\left(-u\frac{\partial T_n}{\partial x} + \frac{k}{\rho c}\frac{\partial^2 T_n}{\partial x^2}\right)\right.$$

$$\left. + \frac{k}{\rho c}\frac{\partial^2}{\partial x^2}\left(\frac{T_{n+1} - T_n}{\Delta t}\right)\right] \tag{6.5.6}$$

Collecting the coefficients of T_{n+1} and T_n, after multiplying throughout by ρc, we have

$$\left(\frac{\rho c}{\Delta t} - \frac{k}{2}\frac{\partial^2}{\partial x^2}\right)T_{n+1} = T_n\left[-\rho c u\frac{\partial}{\partial x} + k\frac{\partial^2}{\partial x^2} + \frac{\Delta t}{2}\rho c u^2\frac{\partial^2}{\partial x^2}\right.$$

$$\left. - \frac{k}{2}\frac{\partial^2}{\partial x^2} + \frac{\rho c}{\Delta t} - \underbrace{\frac{\Delta t}{2}uk\frac{\partial^3}{\partial x^3}}_{K}\right] \tag{6.5.7}$$

To obtain a fully discrete equation, we apply the Galerkin formulation with local approximations of the form

$$T_n(x) = \sum N_i(x)T_i^n \tag{6.5.8}$$

After using Green's theorem on the second-order terms, we get

$$\left(\frac{\mathbf{M}}{\Delta t} + \frac{\mathbf{k}_d}{2}\right)\mathbf{T}_{n+1} = \left(\frac{\mathbf{M}}{\Delta t} - \frac{\mathbf{k}_d}{2} - \rho c(\mathbf{K}_a + \mathbf{K}_{bd})\right)\mathbf{T}_n \tag{6.5.9}$$

where,

$$\mathbf{M} = \int_\Omega \rho c N_i N_j \tag{6.5.10}$$

$$\mathbf{k}_d = \int k \frac{\partial N_i}{\partial x}\frac{\partial N_j}{\partial x} \tag{6.5.11}$$

$$\mathbf{K}_a = \int_\Omega \rho c u N_i \frac{\partial N_i}{\partial x} \tag{6.5.12}$$

$$\mathbf{K}_{bd} = \frac{\Delta t}{2}\int_\Omega \rho c u^2 \frac{\partial N_i}{\partial x}\frac{\partial N_j}{\partial x} \tag{6.5.13}$$

6.6 VELOCITY–PRESSURE–TEMPERATURE FORMULATION (MIXED INTERPOLATION APPROACH) FOR CONVECTION HEAT TRANSFER PROBLEMS

The velocity–pressure formulation is also known as the primitive variables approach because they involve the basic variables in the analysis. The primitive variables approach is the preferred method by several authors since it is the most straightforward finite element procedure for the solution of the non-linear Navier–Stokes equations. The following reasons are given in favour of the velocity pressure temperature formulation.

(1) Only $C°$ continuity is required of the element interpolation functions.

(2) Pressure, velocity, temperature, velocity gradient, temperature gradient (heat flux) and stress boundary conditions can be directly incorporated into the matrix equations.

(3) The formulation can be extended to three-dimensional cases.

(4) Free surface problems are tractable.

The conventional Galerkin approach, which employs weighting functions equal to the interpolation functions, is employed here to derive the element equations. For a general element within a two-dimensional flow domain Ω bounded by curve C, we select u, v, p and T as nodal variables and interpolate these as follows:

$$u^e = \sum N_i(x, y)u_i = \mathbf{Nu} \tag{6.6.1a}$$

$$v^e = \sum N_i(x, y)v_i = \mathbf{Nv} \tag{6.6.1b}$$

$$p^e = \sum N_{pi}(x, y)P_i = \mathbf{Np} \tag{6.6.1c}$$

$$T^e = \sum N_i(x, y)T_i = \mathbf{NT} \tag{6.6.1d}$$

where N_i and N_{pi} are interpolation functions and N_{pi} is one order less than N_i [16–19]. Thus there are more velocity and temperature unknowns than pressure unknowns. The derivation here follows the work of Gartling and Nickell [20]. The Galerkin criteria applied to the two-dimensional equations (6.2.1)–(6.2.4) lead to

$$\int_\Omega \rho\left(\frac{\partial u}{\partial t} + u\frac{\partial u}{\partial x} + v\frac{\partial u}{\partial y} - \frac{\partial}{\partial x}(\sigma_x - p) + \rho\beta g_x(T - T_\infty) - \frac{\partial \tau_{xy}}{\partial y}\right)N_i\,d\Omega = 0 \tag{6.6.2a}$$

$$\int_\Omega \rho\left(\frac{\partial v}{\partial t} + u\frac{\partial v}{\partial x} + v\frac{\partial v}{\partial y} - \frac{\partial}{\partial y}(\sigma_y - p) + \rho\beta g_y(T - T_\infty) - \frac{\partial \tau_{yx}}{\partial x}\right)N_i\,d\Omega = 0 \tag{6.6.2b}$$

$$\int_\Omega \left(\frac{\partial u}{\partial x} + \frac{\partial v}{\partial y}\right)N_{pi}\,d\Omega = 0 \tag{6.6.2c}$$

$$\int_{\Omega}\left(\rho c\left(\frac{\partial T}{\partial t}+u\frac{\partial T}{\partial x}+v\frac{\partial T}{\partial y}\right)-\frac{\partial}{\partial x}\left(K_x\frac{\partial T}{\partial x}\right)-\frac{\partial}{\partial y}\left(K_y\frac{\partial T}{\partial y}\right)\right)N_i\,d\Omega=0$$

(6.6.2d)

The boundary conditions are

$$u=g(x,y)\qquad v=h(x,y)\quad \text{on } S_1 \tag{6.6.3a}$$

The surface tractions $\bar\sigma_x$ and $\bar\sigma_y$ are given by

$$\left.\begin{aligned}\bar\sigma_x&=(\sigma_x-p)n_x+\tau_{xy}n_y\\ \bar\sigma_y&=\tau_{xy}n_x+(\sigma_y-p)n_y\end{aligned}\right\}\text{on }S_2$$

(6.6.3b)
(6.6.3c)

$$T=f(x,y)\quad\text{on }S_3 \tag{6.6.3d}$$

$$-K_x\frac{\partial T}{\partial x}n_x-K_y\frac{\partial T}{\partial y}n_y=q+h(T-T_\infty)\Rightarrow q_in=m(x,y) \tag{6.6.3e}$$

The final equations can be expressed in the following matrix form:

$$
\begin{array}{c}r\\r\\s\end{array}
\begin{bmatrix}\mathbf{M}&0&0\\0&\mathbf{M}&0\\0&0&0\end{bmatrix}
\begin{Bmatrix}\dot u_1\\\dot u_2\\\vdots\\\dot u_r\\\dot v_1\\\dot v_2\\\vdots\\\dot v_r\\\dot p_1\\\dot p_2\\\vdots\\\dot p_s\end{Bmatrix}
+
\begin{array}{c}r\\r\\s\end{array}
\begin{bmatrix}\mathbf{C}+2\mathbf{K}_{11}+\mathbf{K}_{22}&\mathbf{K}_{12}&\mathbf{L}_1\\\mathbf{K}_{12}^{\mathrm T}&\mathbf{C}+\mathbf{K}_{11}+2\mathbf{K}_{22}&\mathbf{L}_2\\\mathbf{L}_1^{\mathrm T}&\mathbf{L}_2^{\mathrm T}&0\end{bmatrix}
\begin{Bmatrix}u_1\\u_2\\\vdots\\u_r\\v_1\\v_2\\\vdots\\v_r\\p_1\\p_2\\\vdots\\p_s\end{Bmatrix}
$$

$$
=\begin{Bmatrix}f_{u_1}\\f_{u_2}\\\vdots\\f_{u_r}\\f_{v_1}\\f_{v_2}\\\vdots\\f_{v_r}\\0\\0\\0\\\vdots\\0\end{Bmatrix} \tag{6.6.4a}
$$

where r is the number of nodes where the velocity components u and v are specified in the element, and s is the number of nodes where the pressure is specified in the element. Note that $r > s$. For an eight-noded isoparametric element, $r = 8$ and $s = 4$.

$$C_n \begin{Bmatrix} \dot{T}_1 \\ \dot{T}_2 \\ \vdots \\ \dot{T}_r \end{Bmatrix} + \{K_c + K_n + K_v\} \begin{Bmatrix} T_1 \\ T_2 \\ \vdots \\ T_r \end{Bmatrix} = \begin{Bmatrix} f_{q_1} + f_{n_1} \\ f_{q_2} + f_{n_2} \\ \vdots \\ f_{q_r} + f_{n_r} \end{Bmatrix} \tag{6.6.4b}$$

where

$$M_w = \int_\Omega \rho N_i N_j \, d\Omega \tag{6.6.5a}$$

$$C_{ij} = \int_\Omega \rho N_i \left(u \frac{\partial N_j}{\partial x} + v \frac{\partial N_j}{\partial y} \right) d\Omega \tag{6.6.5b}$$

$$K_{11_{ij}} = \int_\Omega \mu \frac{\partial N_i}{\partial x} \frac{\partial N_j}{\partial x} \, d\Omega \tag{6.6.5c}$$

$$K_{22_{ij}} = \int_\Omega \mu \frac{\partial N_i}{\partial y} \frac{\partial N_j}{\partial y} \, d\Omega \tag{6.6.5d}$$

$$K_{12_{ij}} = \int_\Omega \mu \frac{\partial N_i}{\partial y} \frac{\partial N_j}{\partial x} \, d\Omega \tag{6.6.5e}$$

$$L_1 = - \int_\Omega N_i^P \frac{\partial N_i}{\partial x} \, d\Omega \tag{6.6.5f}$$

$$L_2 = - \int_\Omega N_i^P \left\{ \frac{\partial N_i}{\partial y} \right\} d\Omega \tag{6.6.5g}$$

$$f_{ui} = \int_\Omega \bar{\sigma}_x N_i \, dS_2 - \int_\Omega \rho \beta g_x (T - T_\infty) \, d\Omega \tag{6.6.5h}$$

$$f_{vi} = \int_\Omega \bar{\sigma}_y N_i \, dS_2 - \int_\Omega \rho \beta g_y (T - T_\infty) \, d\Omega \tag{6.6.5i}$$

$$C_{n_{ij}} = \int_\Omega \rho c N_i N_j \, d\Omega \tag{6.6.5j}$$

$$K_{c_{ij}} = \int_\Omega \left(K_x \frac{\partial N_i}{\partial x} \frac{\partial N_j}{\partial x} + K_y \frac{\partial N_i}{\partial y} \frac{\partial N_j}{\partial y} \right) d\Omega \tag{6.6.5k}$$

$$K_{n_{ij}} = \int_S h N_i N_j \, dS \tag{6.6.5l}$$

$$K_{v_{ij}} = \int_\Omega \rho c \left(u N_i \frac{\partial N_j}{\partial x} + v N_i \frac{\partial N_j}{\partial y} \right) d\Omega \tag{6.6.5m}$$

$$f_{q_i} = \int_S - qN_i \, dS \tag{6.6.5n}$$

$$f_{h_i} = \int_S hT_\infty N_i \, dS \tag{6.6.5p}$$

If there is heat generation, G, within the domain, then $\{f_G\}$ is to be added to the force term

$$f_{G_i} = \int_\Omega GN_i \, d\Omega \tag{6.6.5q}$$

The coefficient matrices $[M]$ and $[C_n]$ of the unsteady velocity components and temperature are the element mass matrix and capacitance matrix, respectively. The mass and capacitance matrices calculated based on equations (6.6.5a) and (6.6.5j), respectively, are called consistent matrices. As an approximation we can use the 'lumped' approach, thus making the matrix diagonal. The diagonalisation of the matrix permits the use of an explicit transient time integration algorithm like Euler's explicit method. Gresho *et al.* [21] discuss the effect of mass lumping in advection-dominated flows. The diffusive terms like K_{11} and K_{22} and K_c are symmetric in nature while the convective matrices $[C]$ and $[K_v]$ are asymmetric in nature, which represent convection of momentum and heat, respectively. The element equations are then assembled and the boundary conditions subsequently imposed. The resulting system of non-linear equations is solved by iteration. If the forced convection problem is being solved, the hydrodynamic part can be solved independently of the energy equation and consequently the temperature can be determined. However, in the cases of free convection or forced convection with temperature-dependent properties, both the hydrodynamic and the thermal part have to be solved simultaneously. For unsteady convection problems, a set of non-linear ordinary differential equations has to be solved by using a time integration scheme with an iteration scheme at each time step. Proper element choice, effective solution algorithm, correct use of boundary conditions and an appropriate discretisation of the solution domain are essential for a good solution [4, 22, 23].

6.7 EXAMPLES OF HEAT TRANSFER IN A FLUID FLOWING BETWEEN PARALLEL PLANES

The earliest work pertaining to the application of the finite element method to convection heat transfer problems was made by Tay and De Vahl Davis [24]. The problem they considered required the determination of the temperature distribution and the axial variation of the local Nusselt number for a fluid of constant physical properties flowing between two infinite parallel planes. The temperature and heat fluxes are specified at the surfaces and the flow is assumed

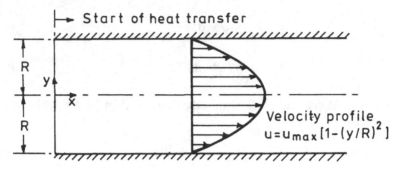

Figure 6.7.1 *Fully developed flow between two parallel planes. (Reproduced with permission from Tay A.O. and De Vahl Davies G., Application of the finite element method to convection heat transfer between parallel plates. Int. J. Heat Mass Transfer, **14**, 1057–1069 (1971), © Pergamon Press.)*

to be hydrodynamically developed (parabolic), as shown in Figure 6.7.1. For the problem under consideration, the governing differential equation is

$$\rho c_p u \frac{\partial T}{\partial x} = k \left[\frac{\partial^2 T}{\partial x^2} + \frac{\partial^2 T}{\partial y^2} \right] \tag{6.7.1}$$

with the boundary conditions

$$T = T_s \qquad \text{on surface } S_1 \tag{6.7.1a}$$

$$-k \nabla T = q \quad \text{on surface } S_2 \tag{6.7.1b}$$

The following set of dimensionless variables is introduced:

$$X = \frac{2x}{D_n P_e} \qquad Y = \frac{y}{R} \qquad u^+ = \frac{u}{u_{max}} = 1 - Y^2 \qquad \theta = \frac{T - T_e}{\varepsilon}$$

and hence the given equation reduces to

$$u^+ \frac{\partial \theta}{\partial x} = 8 \left[\frac{1}{(2P_e)^2} \frac{\partial^2 \theta}{\partial x^2} + \frac{\partial^2 \theta}{\partial y^2} \right] \tag{6.7.2}$$

where $\varepsilon = C$ for the case of a linearly varying wall temperature $T_w = T_e + CX$ and $\varepsilon = -q_w D_h / K$ for a constant heat flux at the upper wall and insulated at the lower wall.

A variational formulation is used to determine the element characteristics. Linear triangular elements are utilised in the solution. For the case where both top and bottom surfaces are subjected to uniformly varying temperature, only the upper half is considered for the solution, as shown in Figure 6.7.2. The elements near the wall and the entrance are small since it is expected that the temperature gradients will be greater in these regions. The length of the mesh is chosen to be sufficiently long such that the right-hand boundary is in the

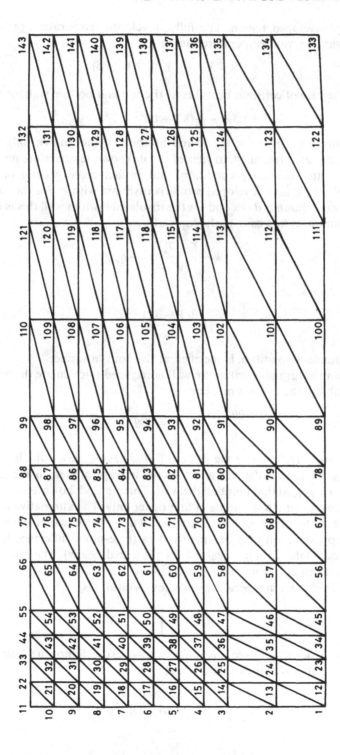

igure 6.7.2   *A typical 13 × 11 mesh. (Reproduced with permission from Tay A.O. and De Vahl Davies G., Application of the finite element ethod to convection heat transfer between parallel plates. Int. J. Heat Mass Transfer, **14**, 1057–1069 (1971), © Pergamon Press.)*

thermally developed region. The fully developed temperature profile for the linearly varying wall temperature case is

$$\theta = X - \tfrac{1}{96}(Y^4 - 6Y^2 + 5)$$

and for the case of constant heat flux at the top and bottom wall insulated.

$$\theta = 1.5X + \tfrac{1}{84}(8Y + 6Y^2 - Y^4) - \tfrac{39}{2240}$$

At the entrance, $\theta = 0$, for all Y, and at the exit the known fully developed temperatures are prescribed. In view of the unknown temperature profiles at the exit, the better boundary condition, and in fact more general in nature, is $\partial\theta/\partial x = 0$, i.e. the fully developed profile is invariant with x. For the top wall, the boundary condition is $\theta = x$ and for the insulated bottom wall this is left free. In the constant heat flux case, at the top wall:

$$k\frac{\partial(T - T_e)}{\partial y} = -q_w;$$

at the bottom wall:

$$k\frac{\partial(T - T_e)}{\partial y} = 0$$

i.e. no boundary condition is specified at the bottom surface.

The temperature gradient at the wall is required to calculate the heat transfer coefficient and the Nusselt number:

$$\frac{\partial\theta}{\partial y} = \frac{1}{2\Delta}(C_i\theta_i + C_j\theta_j + C_m\theta_m)$$

Hence we may calculate the values of $\partial\theta/\partial y$ associated with all the elements adjoining the node at the wall and take the average of these as the temperature gradient $(\partial\theta/\partial y)_w$ at that point on the wall. This method, however, is not accurate since the approximation will be of first order only. The alternative method is to employ a finite difference formula, which will be more accurate since a higher order approximation may be employed (a three-point difference formula was used to calculate the numerical results presented below). The same comments apply to the calculation of $\partial\theta/\partial x$ and $\partial^2\theta/\partial x^2$.

The mixed mean temperature (bulk temperature) is

$$\theta_{mm} = \frac{3}{4}\int_{-1}^{1} u^+ \, dY$$

Integration across the section is performed using Simpson's rule. The local Nusselt number Nu_x is calculated from

$$Nu_x = \frac{4(\partial\theta/\partial y)}{\theta_w - \theta_{mm}}$$

The numerical results are then compared with the theoretical solutions available in [25]. In the theoretical analysis the axial conduction term is neglected i.e. $\partial^2\theta/\partial x^2 = 0$. This can easily be incorporated into a finite element analysis once the value of P_e is specified.

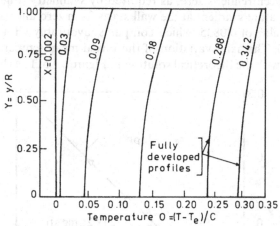

Figure 6.7.3 *Temperature profile for fluid for linearly varying wall temperature. (Reproduced with permission from Tay A.O. and De Vahl Davies G., Application of the finite element method to convection heat transfer between parallel plates. Int. J. Heat Mass Transfer,* **14**, *1057–1069 (1971), © Pergamon Press.)*

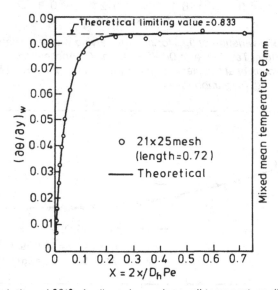

Figure 6.7.4 *Variation of ∂θ/∂y for linearly varying wall temperature. (Reproduced with permission from Tay A.O. and De Vahl Davies G., Application of the finite element method to convection heat transfer·between parallel plates. Int. J. Heat Mass Transfer,* **14**, *1057–1069 (1971), © Pergamon Press.)*

Figure 6.7.3 shows the temperature profiles for the case of both walls varying linearly. At small values of x, virtually no heat is transported to the central region and hence the temperature remains constant. For larger values of x the temperature profiles gradually become fully developed. The transverse temperature gradient at the centreline is zero, as required by symmetry. Figure 6.7.4 shows that the temperature gradient at the wall starts from zero and rapidly increases to a limiting value of 0.0835, which compares favourably with the theoretical value of 0.0833. The axial variation of the mixed mean temperature is plotted and compared with the theoretical solution in Figure 6.7.5. The slope of θ_m versus

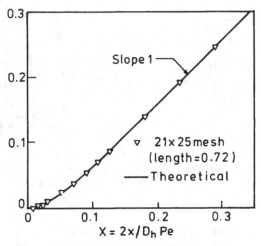

Figure 6.7.5 *Axial variation of θ_{mm} for linearly varying wall temperature. (Reproduced with permission from Tay A.O. and De Vahl Davies G., Application of the finite element method to convection heat transfer between parallel plates. Int. J. Heat Mass Transfer, **14**, 1057–1069 (1971), © Pergamon Press.)*

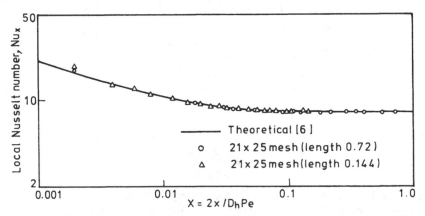

Figure 6.7.6 *Variation of Nu_x for both walls with linearly varying temperature. (Reproduced with permission from Tay A.O. and De Vahl Davies G., Application of the finite element method to convection heat transfer between parallel plates. Int. J. Heat Mass Transfer. **14**. 1057–1069 (1971). © Pergamon Press.)*

X curve leads to a value of 1 as expected, since in the fully developed region $\partial\theta/\partial X = 1$ for all Y.

The computed variation of the local Nusselt number, Nu_x, in the direction of the flow is plotted and compared with the theoretical solution in Figure 6.7.6. Agreement is very good except near the entrance. The Nusselt number at the entrance section is of course infinite but rapidly decreases to a limiting value of 8.24. This value of 8.24 occurs in the numerical solution and confirms that the

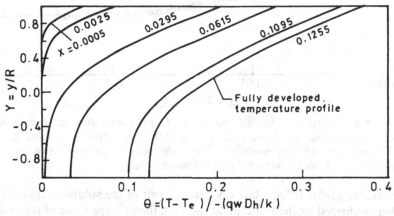

Figure 6.7.7   *Temperature profile in the fluid for upper wall constant heat flux. (Reproduced with permission from Tay A.O. and De Vahl Davies G., Application of the finite element method to convection heat transfer between parallel plates. Int. J. Heat Mass Transfer, **14**, 1057–1069 (1971), © Pergamon Press.)*

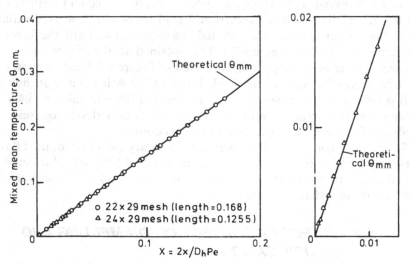

Figure 6.7.8   *Axial variation of θ_{mm} for upper wall constant heat flux. (Reproduced with permission from Tay A.O. and De Vahl Davies G., Application of the finite element method to convection heat transfer between parallel plates. Int. J. Heat Mass Transfer, **14**, 1057–1069 (1971), © Pergamon Press.)*

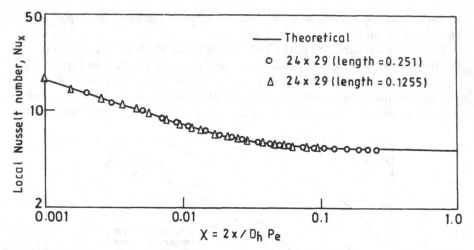

Figure 6.7.9 *Variation of Nu_x for the case of upper wall with constant heat flux. (Reproduced with permission from Tay A.O. and De Vahl Davies G., Application of the finite element method to convection heat transfer between parallel plates. Int. J. Heat Mass Transfer, **14**, 1057–1069 (1971), © Pergamon Press.)*

temperature profile is fully developed at the exit of the solution region. This is further confirmed by the fact that, at the exit, the average value of $\partial\theta/\partial x$ across the channel has risen to almost 1.0 and that of $\partial^2\theta/\partial x^2$ has fallen to almost zero.

The second case considered is that where the upper wall is subjected to a constant heat flux and the lower wall is insulated. The temperature profiles are shown in Figure 6.7.7. As expected, the value of $\partial\theta/\partial y$ is zero at the lower wall because it is insulated. All along the upper wall, $\partial\theta/\partial y$ is equal to within 1% of the theoretical dimensionless temperature gradient of 0.25. The axial variation of the mixed mean temperature is plotted and compares well with the theoretical solution $\theta_{mm} = 1.5X$ in Figure 6.7.8. The obtained axial variation of the local Nusselt number of the upper wall is plotted in Figure 6.7.9 and compares well with the theoretical values from [25]. The local Nusselt number at the upper wall is infinite at the entrance but decreases to a limiting value of 5.40. This implies that the temperature profile at the exit is fully developed (since also $\partial\theta/\partial x = 1.5$ and $\partial^2\theta/\partial x^2 = 0$ across the exit section).

The case of a constant wall temperature boundary condition for fluid flowing between parallel plates is studied by Ravikumaur *et al.* [26], who also considered the effect of a triangular step located on one of the walls on the pressure drop and heat transfer characteristics.

6.8 EFFECT OF CONVECTION ON MELTING AND SOLIDIFICATION

The phase change problems have been analysed using conduction heat transfer only, since this is sufficient for many cases; however, many analyses have appeared taking into account natural convection effects. Convection cannot be neglected in many cases of practical interest [27]. A significant contribution is made by Gartling [28] who made use of the Boussinesq approximation and the enthalpy method in solving the Navier–Stokes and energy equations. Later Morgan [29] presented an explicit finite element algorithm for the solution of the basic equations describing conductive and convective transfer of heat in a material undergoing liquid/solid change of phase in a cylindrical thermal cavity. Recently Usmani *et al.* [30] have proposed a model based on the treatment of release/absorption of latent heat as a heat source/sink in combination with the standard Galerkin finite element method with a primitive variable formulation of a fixed grid. Three cases of phase change of an aluminium alloy in the presence of natural convection are considered, i.e. solidification, melting and combined solidification and melting. The solidification of water in a square cavity is modelled as another example, taking into account the density extremum, and

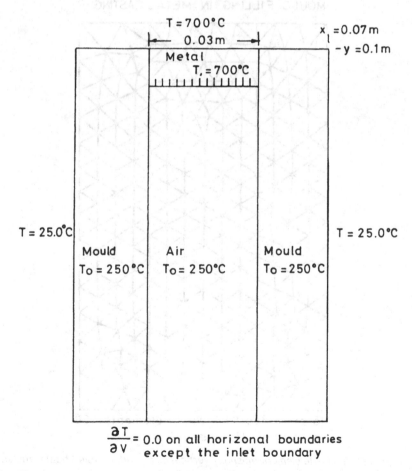

Figure 6.9.1 *Configuration for vertical filling with coupled heat transfer.*

the results compare well with the numerical results of Davis *et al.* [31] and experimental results of Weaver and Viskanta [32].

6.9 MOULD FILLING IN CASTINGS

Although the critical part of a casting analysis must obviously be the cooling and solidification stage, the conditions under which the mould fills can also have an important influence over the quality of the finished casting. Moreover, the design of the filling system is an important part of the die designer's work, and is often the area in which he has the greatest latitude for design changes. The primary benefit of a mould filling simulation is that it will provide a more realistic, non-uniform initial temperature distribution for the solidification stage and will

MOULD FILLING IN METAL CASTING

No. of Nodes:985 No. of elements:468
No. of Nodes/Element :6

Figure 6.9.2   *Finite element mesh for vertical filling with coupled heat transfer.*

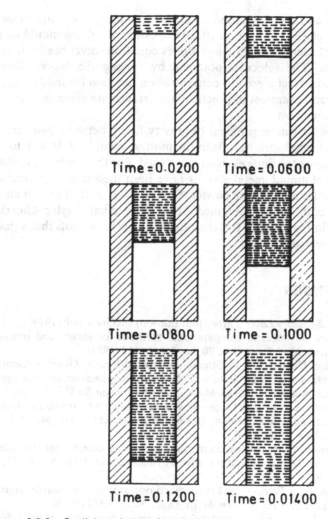

Time = 0.0200   Time = 0.0600

Time = 0.0800   Time = 0.1000

Time = 0.1200   Time = 0.01400

Figure 6.9.3 *Explicit solution for vertical filling with coupled heat transfer: front positions.*

also provide an initial velocity field for the modelling of convection. Examination of the flow patterns during filling can reveal a wide range of potential difficulties with the height, dimensions and positions of risers or the gating system. By including the phase change effects during filling it is possible to predict freeze-off, where the flow through a narrow section freezes before reaching its intended destination, and cold shuts, where molten metal meets already solidified metal, causing a fault line. With even simple dies costing large sums of money the ability to expose flaws in a die design before the die is made is a valuable asset.

Gao *et al.* [33] have tackled the mould filling problems in metal castings for horizontal, vertical and a combination of these cases using the finite element method. Usmani *et al.* [34] provide an effective method of solving mould filling problems. A finite-element-based Navier–Stokes equation solver has been used to analyse the flow. The velocities obtained by solving the Navier–Stokes equation are used to advect a pseudo-concentration function for modelling the free fluid front. Several examples of practical importance are given to illustrate the mould filling simulation.

Figure 6.9.1 shows a simple problem of gravity filling between two vertical walls of a mould, which absorb heat from the molten metal as it flows into the void. Figure 6.9.2 shows the finite element mesh used. For this problem, realistic thermal properties of mould, metal and air have been used in the appropriate regions. The properties used for the Navier–Stokes solver correspond to an *Re* of approximately 20.0. Results obtained from the explicit Taylor–Galerkin analysis are shown in Figure 6.9.3. It is clear from the isotherm plots that a sharp thermal front is advected faithfully.

REFERENCES

[1] Schlichting H., *Boundary Layer Theory* (6th Edition) McGraw-Hill (1979).
[2] Donea J., Recent advances in computational methods for steady and transient transport problems *Nucl. Engg. Des.*, **80**, no. 2, 141–162 (1984).
[3] Brooks A. N. and Hughes T. J. R., Streamline upwind/Petrov–Galerkin formulations for convection dominated flows with particular emphasis on the incompressible Navier–Stokes equations *Comp. Meth. Appl. Mech. Engg*, **32**, 199–259 (1982).
[4] Gresho P. M. and Lee R. L., Don't suppress wiggles — they are telling you something *Finite Element Methods for Convection Dominated Flows*, AMD, Vol. 34, Ed. T.J.R. Hughes ASME, New York (1979).
[5] Allen D. and Southwell R., Relaxation methods applied to determining the motion in two-dimensions of a viscous fluid past a fixed cylinder *Quart. J. Mech. Appl. Math.*, **8**, 129–145 (1955).
[6] De Sampio P. A. B., A Petro–Galerkin/modified operator formulation for convection–diffusion problems *Int. J. Num. Meth. Engg*, **30**, 331–347 (1990).
[7] Hughes T. J. R., A simple scheme for developing upwind finite elements *Int. J. Num. Meth. Engg*, **12**, 1359–1365 (1978).
[8] Donea J., Belytschko T. and Smolinski P., A generalised Galerkin method for steady convection–diffusion problems with applications to quadratic shape function elements *5th Int. Conf. on Finite Elements and Flow Problems*.
[9] Hughes T. J. R. and Brooks A. N., A multi-dimensional upwind scheme with no cross wind diffusion *Finite Element Methods for Convection Dominated Flows*, AMD, Vol. 34, Ed. T.J.R. Hughes ASME, New York (1979).
[10] Donea J., A Taylor–Galerkin method for convective transport problems *Int. J. Num. Meth. Engg*, **20**, 101–119 (1984).
[11] Morton K. W. and Parrott A. K., Generalised Galerkin methods for first order hyperbolic equations *J. Comp. Phys.* **36**, 249–270 (1980).
[12] Roache P. J., *Computational Fluid Mechanics* Hermosa Publishers, Alburqueque, USA (1976).

[13] Lohner R., Morgan K. and Zienkiewicz O. C., The solution of non-linear hyperbolic equation systems by the finite element method, *Int. J. Num. Meth. Fluids*, **4**, 1043–1063 (1984).

[14] Gresho P. M., Chan S. T., Lee R. L. and Upson C. D., A modified finite element method for solving the time dependent incompressible Navier–Stokes equations *Int. J. Num. Meth. Fluids*, **4**, 557–598 (1984).

[15] Donea J., Giuliani S., Laval H. and Quartapelle L., Time accurate solution of advection–diffusion problems by finite elements, *Comp. Meth. Appl. Mech. Engg*, **45**, 123–145 (1984).

[16] Yamada Y., Ito K., Yokouchi Y., Tamano T. and Ohtsubo T., *Finite Element Analysis of Steady Fluid and Metal Flow*, Vol. 1, Eds. R. H. Gallaghar, J. T. Oden, C. Taylor and O. C. Zienkiewicz John Wiley & Sons Ltd, Chapter 4, pp. 73–93 (1975).

[17] Hood P. and Taylor C., Navier Stokes equations using mixed interpolation *Proc. of First Int. Symp. on Finite Elements Methods in Flow Problems*, Univ. College of Swansea, Wales, pp. 121–132 (1974)

[18] Olson M. D. and Tuann S. Y., Primitive variables versus stream function finite element solutions of the Navier–Stokes equations *Finite Element in Fluids*, Vol. 3, Eds. R. H. Gallagher *et al.*, pp. 77–87 John Wiley & Sons Ltd (1978).

[19] Bercovier M. and Piromean O., Error estimates for finite element method solution of the Stokes problems in primitive variables *Num. Math.*, **33**, 211–224 (1979).

[20] Gartling D. K. and Nickell R. E., Finite element analysis of free and forced convection *Finite Elements in Fluids*, Vol. 3, Eds. R. H. Gallagher *et al.*, Chapter 6, pp. 105–121 John Wiley & Sons Ltd (1978).

[21] Gresho P., Lee R. L. and Sani R., Advection dominated flows with emphasis on the consequence of mass lumping *Finite Elements in Fluids*, Vol. 3, Eds. R.H. Gallagher *et al.*, Chapter 19, pp. 335–350, John Wiley & Sons Ltd (1978).

[22] Gartling D. K., Nickell R. E. and Tanner R. I., A finite element convergence study for accelerating flow *Int. J. Num. Meth. Engg*, **11**, 1155–1174 (1977).

[23] Ben-sabar E. and Caswell B., A stable finite element simulation of convective transport *Int. J. Num. Meth. Engg*, **14**, 545–565 (1979).

[24] Tay A. O. and De Vahl Davis G., Application of the finite element method to convection heat transfer between parallel plates *Int. J. Heat Mass Transfer*, **14**, 1057–1069 (1971).

[25] Lundburg R. E., Reynolds W. C. and Kays W. M., *Heat Transfer With Laminar Flow in Concentric Annuli with Constant and Variable Wall Temperature and Heat Flux* NASA TN D-1972, Washington, DC, August (1963).

[26] Ravikumaur S. G., Aswatha Narayana P. A., Seetharamu K. N. and Ramaswamy B., Laminar flow and heat transfer over a two-dimensional triangular steps *Int. J. Num. Meth. Fluids*, **9**, 1165–1177 (1989).

[27] Viskanta R., Natural convection in melting and solidification *Natural Convection; Fundamentals and Applications*, Eds. S. Kakac *et al.* Hemisphere, Washington, D.C., pp. 845–877 (1985).

[28] Gartling D. K., Finite element analysis of convective heat transfer problems with change of phase *Computer Methods in Fluids*, Eds. K. Morgan *et al.* Pentech, London, pp. 489–500 (1980).

[29] Morgan K., A numerical analysis of freezing and melting with convection *Comp. Meth. Appl. Mech. Engg*, **28**, 275–284 (1981).

[30] Usmani A. S., Lewis R. W. and Seetharamu K. N., Finite element modelling of

natural convection controlled change of phase *Int. J. Num. Meth. Fluids*, **14**, 1019–1036 (1992).

[31] Davis G. V., Leonardi P. H., Wong P. H. and Yeoh G. H., Natural convection in a solidifying liquid, *6th Int. Conf. for Numerical Methods in Thermal Problems* Swansea, pp. 410–420 (1989).

[32] Weaver J.A. and Viskanta R., Freezing of water saturated porous media in a rectangular cavity *Int. Comm. Heat Mass Transfer*, **13**, 245–251 (1986).

[33] Gao D. M., Dhatt G., Belanger J. and Cheikh A. B., A finite element simulation of metal flows in moulds, *6th Int. Conf. for Numerical Methods in Thermal Problems* Swansea, pp. 421–430 (1989).

[34] Usmani A. S., Cross J. T. and Lewis R. W., A finite element model for the simulation of mould filling in metal casting and associated heat transfer, *Int. J. Num. Meth. Engg*, **35**, 787–806 (1992).

7 Further Development

7.1 INTRODUCTION

The successful numerical solution of the range of heat transfer problems described in this book enables a series of further related engineering problems to be tackled. While the major part of this book is devoted to a detailed explanation of heat transfer problems alone, it is the intension of this chapter to illustrate some of these further problems. In particular, the work presented here describes problems that have been solved by the authors in their research work. These can be classified into three broad categories:

- the development of thermal stresses arising out of non-uniform temperature distributions in a material;
- coupled heat and moisture movement in a capillary porous material;
- shrinkage stress development arising out of non-uniform moisture content distributions in a drying material.

Coverage of these topics will inevitably be much briefer than has been provided for the heat transfer work alone but further details can be obtained from the references quoted in the chapter.

7.1.1 Thermal stress development

Thermal stress development is important in a vast range of industrial and commercial applications. The basic physical explanation for the phenomenon is related to the simple fact that as the temperature of a material changes, the material expands or contracts. The well-known coefficient of thermal expansion is used to relate the two processes. The coefficient itself is determined from experiments on samples of the material free to expand or contract in an unrestrained manner. Problems arise, however, when temperature changes take place in more complex structures since in these cases thermal expansion is restrained. Consequently, stresses are developed that are potentially a threat to the integrity of the overall system.

The numerical analysis of the problem is structured to follow the physical stress development as closely as possible. In the first place the unrestrained deformation that would take place in each element of the domain is calculated. This is achieved by multiplying the change in element temperature over a particular time interval by the relevant coefficient of thermal expansion. These unrestrained deformations can be converted to strains and stresses using standard finite element methods [1]. The stresses can then be converted to 'equivalent nodal forces', again using standard finite element techniques. Since it is the structure's inability to allow unrestrained deformation that is the source of stress development in the material, the 'equivalent nodal forces' are applied to the problem to restore its deformations to its original state. These 'equivalent nodal forces' then act, of course, as real loading terms in the problem and deformations and stresses from this loading can be predicted by means of a suitable finite element stress analysis. These deformations and stresses are the real deformations and stresses that would take place in the problem under consideration. This approach is known as the 'initial strain' method and is well documented in the literature [1].

The quality of the results achieved depends on the correct representation of the material's stress/strain relationship in the numerical stress analysis model. For some materials a linear elastic stress/strain relationship adequately describes behaviour and this case will be described here [2]. For other materials more complex models will be required and to this end an elasto-plastic stress/strain model will also be included [3]. The elasto-plastic solution is in fact achieved using an elasto/viscoplastic [4] numerical model, operated in a mode whereby the viscosity of the material is used as an artifice to achieve the true elasto-plastic results.

A solution of thermal stress development is therefore achieved using thermal and stress numerical analyses operating in tandem. Results from the appropriate heat transfer analysis are fed through to the appropriate stress analysis. Any coupling of the dependence of heat transfer on stress development is assumed to be insignificant compared with the dependence of stress development on heat transfer and is consequently ignored. The time-varying stress development pattern is achieved by marching forward in time, with heat transfer results from each time step being fed into the stress analysis for that particular time step. The whole approach is demonstrated in this chapter for the case of stress development during the solidification of metal bars, strands and ingots.

7.1.2 *Heat and moisture transfer*

Coupled heat and moisture movement problems are also important in a range of engineering applications. The treatment in this book examines the particular case of coupled heat and moisture transfer in capillary porous materials, which is itself a problem of interest to virtually the whole spectrum of engineering disciplines. A capillary porous material consists of three distinct phases: solid particles which form the matrix of the material, liquid and air. The air in turn may contain vapour emanating from the liquid in the matrix. When a temperature gradient is imposed across such a material a number of phenomena are initiated which are both complex in themselves and also interrelated. For example, an imposed temperature gradient causes moisture flow to take place in both the liquid and the vapour phase. A moisture content gradient is thus established that will cause further moisture flow to take place. Changes in moisture content then cause changes in the physical characteristics of the material and the thermophysical parameters which determine heat and moisture transfer rates are changed. This further complicates the problem and can lead to the so-called 'thermal instability' [5] phenomena.

Heat transfer in a capillary porous body is no longer confined only to conduction. Other factors now also need to be included. Heat is consumed in capillary porous materials in the vaporisation of water. Consequently, when moisture transfer takes place in the form of vapour the component of latent heat expended in the vaporisation of that vapour is also transferred. This must be included in the heat balance equations. Furthermore, in some capillary porous materials, where the permeabilities are high, sensible heat may be transferred by bulk flow of liquid and vapour. In other words convection effects also have to be included.

During an intense period of heating of the material, such as high frequency convective drying, a gradient of total pressure arises by the evaporation of liquid at a temperature greater than that of the boiling point of water. The presence of the total pressure gradient causes a hydrodynamic or filtration transfer of liquid within the porous media. The pressure of the entrapped air bubbles, contained within the pores of the material, is increased upon heating; hence the air bubbles expand, forcing the liquid and vapour along in the flow of the direction of heat. This is known as filtration moisture transfer. During a non-intense period of drying the pressure gradient has time to 'dissipate' and does not become significant. During intense drying the filtration transfer of moisture can be significantly greater than the transfer of moisture caused by diffusion or capillarity, and consequently the material to be dried will reach a steady state much quicker because moisture is being transferred to the material surface much faster.

Considerable basic research has been carried out on the problem of heat and mass transfer in capillary-porous bodies and a number of viable theories have been postulated. Krischer [6], for example, has proposed a theory which is based on the assumption that during drying some combination of capillary flow of liquid and diffusion of vapour controls the internal moisture diffusion. A similar approach has been proposed by Philip and De Vries [7] and De Vries [8]. Their model was originally developed for applications in soil physics to soil moisture migration problems. A more rigorous formulation of the problem has been presented by Luikov [9] who developed a theory from the methods of the thermodynamics of irreversible processes. Considerable work was carried out by

Luikov to determine experimentally the values of the parameters required in the theory and also to validate the theory as far as possible. The resulting work appears to model the physical process well. In the absence of any definitive evidence of the superiority of one approach over another, Luikov's model has been chosen as a suitable model for solution.

The model is applied to a problem of interest in the drying industry, namely the drying of green timber. Freshly felled timber has typically a moisture content ranging from 60% to 200%. This must be reduced to a value between 9% and 18% to ensure satisfactory end usage. Drying is necessary to ensure dimensional stability, resistance to decay and increased strength. Although timber has been used as a building material since time immemorial, the drying process and drying schedules are largely empirical. Therefore there is a need for an improved understanding of the way in which drying rates are affected by the various controlling factors that can be applied. This can be achieved using the numerical model proposed.

The timber drying problem is solved numerically using Luikov's two degree of freedom system of partial differential equations which describes the variation of temperature and moisture content in the capillary-porous body. No account is taken of the pressure gradient. The second problem that will be presented examines the moisture content variation in an encapsulated electronic circuit. The electronic circuit container must remain below a specified level of moisture content in order that the electronic circuit is not damaged. This problem is solved using both Luikov's two and three degree of freedom systems and a comparison is made of the results to highlight the influence of the pressure gradient upon the solutions.

7.1.3 Shrinkage stress development

The final section of this chapter considers the problem of shrinkage stresses in a capillary-porous body. Shrinkage occurs when moisture is removed by thermal means. The resulting non-uniformities in moisture content give rise to the development of shrinkage stresses. The problem is therefore analogous to the problem of thermal stress development discussed in this chapter. The method of analysis adopted here is therefore the same, the initial strains in the 'initial strain' approach being calculated now from shrinkage coefficients rather than thermal expansion coefficients. Once again results are achieved using two numerical analyses operating in tandem. In this case, however, the results of a heat and mass transfer analysis are fed through to the appropriate stress analysis program. In a similar way to the heat transfer problem, the dependence of heat and moisture transfer on stress development is assumed to be insignificant compared with the dependence of stress development on moisture transfer.

The correct representation of the material's stress/strain relationship is equally of paramount importance in the evaluation of shrinkage stresses. Elastic [10] and elasto-plastic [11] problems have again been considered and interesting results have been obtained. However, to extend the generality of the formulation a step further, and continue the analysis of the problem considered in the heat and mass transfer section, a stress/strain relationship applicable to anisotropic materials will be described here. Timber is in fact an orthotropic material, its stress/strain relationship being different in the longitudinal, tangential and radial directions. Therefore the particular anisotropic case of orthotropy will be dealt with. Both elastic and elasto-plastic constitutive relationships will again be considered.

A further rheological model of interest in the case of timber drying is that which exhibits visco-elastic behaviour. This stress/strain relationship will also be examined, again for the case of orthotropic behaviour.

7.2 THERMAL STRESS DEVELOPMENTS

The basis of the two constitutive relationships employed in this work are first described in this section. The finite element solution procedure is then explained before proceeding to a presentation of results achieved. The application of the method is demonstrated by means of problems associated with the solidification of metals.

7.2.1 Elastic stress/strain relationships

The linear elastic behaviour of materials is well known and need not be described in detail here. For an isotropic material the stress/strain relationships are usually written as

$$\varepsilon_x = \frac{\sigma_x}{E} - \frac{v\sigma_y}{E} - \frac{v\sigma_z}{E}$$

$$\varepsilon_y = \frac{-v\sigma_x}{E} + \frac{\sigma_y}{E} - \frac{v\sigma_z}{E}$$

$$\varepsilon_z = \frac{v\sigma_x}{E} - \frac{v\sigma_y}{E} + \frac{\sigma_z}{E}$$

(7.2.1)

$$\gamma_{xy} = \frac{1}{G}\tau_{xy}$$

where E is Young's modulus, v Poisson's ratio, G the shear modulus and the three are related by the expression

$$G = \frac{E}{2(1 + v)}$$

(7.2.2)

In finite element notation these equations can be written in matrix form as

$$\varepsilon^e = D^{-1}\sigma \tag{7.2.3}$$

7.2.2 Elasto/viscoplastic rheological model

The rheological behaviour of many engineering materials is complex and consists of a combination of time-dependent phenomena such as creep and progressive failure, together with time-independent elastic or plastic straining. The rheologic effects are usually more pronounced after the plastic state has been reached. An elasto/viscoplastic model is an attempt to represent this behaviour by combining a linear elastic model with a plastic analysis which contains not only plastic straining but also time-dependent behaviour.

The concept of viscoplasticity is best illustrated by means of simple uniaxial rheological models. The components that combine to form these models are shown individually in Figure 7.2.1, together with their stress/strain behaviour. Pure elasticity is represented by a Hookean spring which has a linear relationship between stress and strain, pure plasticity is represented by a Saint Venant's slider which has zero strain until the stress reaches the yield stress, σ_{yield}, at which point the strains become indeterminate, and the viscous component is represented by a Newtonian dashpot in which the strain rate is related to the applied stress by the viscosity.

The elasto/viscoplastic rheological model consists of a slider and a dashpot in parallel, and this combination is in series with a Hookean spring. For the uniaxial model shown in Figure 7.2.2 the behaviour is purely elastic until the stress level exceeds the yield stress, σ_{yield}. Once in the plastic region the viscous component becomes active and for rapidly applied loads the stresses can exceed the plastic limit. If unloading takes place from the yielded state the strain path followed is different from that of loading and permanent deformation takes place. Elasto/viscoplastic behaviour is also loading path dependent since it has been demonstrated by experiments that if two loading paths reach the same point on the yield surface by different routes, then the plastic strains are different. This necessitates the use of an incremental form of plasticity solution in which the increments of plastic strain throughout the loading history are determined and the total strain obtained by summation.

Once the elastic limit has been reached for an elasto/viscoplastic material the total strain ε_{ij}, can be expressed as the sum of the elastic ε_{ij}^e and inelastic ε_{ij}^{vp} strain, where the inelastic strain represents the combined effects of viscous and plastic components, i.e.

$$\varepsilon_{ij} = \varepsilon_{ij}^e + \varepsilon_{ij}^{vp} \tag{7.2.4}$$

In the case where some form of initial strain was already locked into the body

$$\varepsilon_{ij} = \varepsilon_{ij}^e + \varepsilon_{ij}^{vp} + \varepsilon_{ij}^i \tag{7.2.5}$$

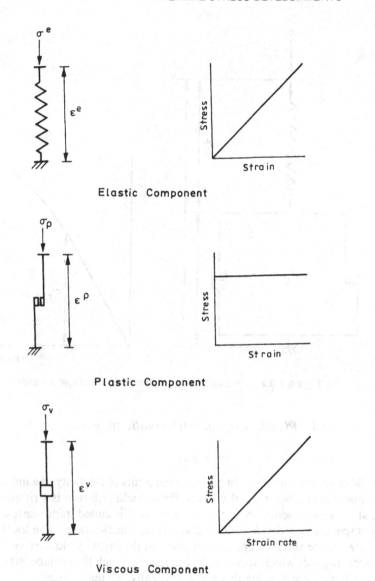

Figure 7.2.1 *Elastic plastic and viscous rheological models.*

Thus, in order to obtain a solution using an elasto/viscoplastic model, the following general relations between stress and strain must be specified:

(1) Elastic stress/strain relationships as described in Section 7.2.1.

(2) The stress conditions that indicate the onset of plastic flow and an end of the elastic behaviour (a yield criterion).

(3) A relationship to determine the plastic strain.

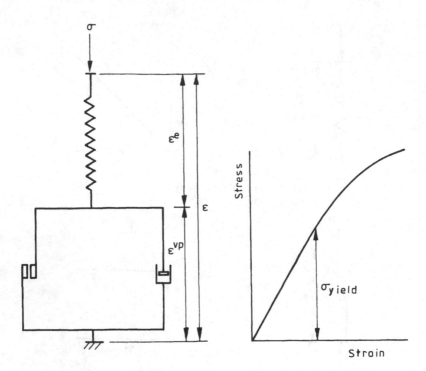

Figure 7.2.2   *Uniaxial elasto/viscoplastic rheological model.*

7.2.3   Plastic stress/strain relationships

7.2.3.1   The yield criterion

A yield criterion is an attempt to define the limits of elasticity of a material, under any possible combination of stresses. If this yield criterion is written in terms of the stress components, the yield function, as it is called, represents a surface in stress space which divides elastic states from plastic states. The location, shape and size of the surface depends not only on the initial yield surface but also on a hardening rule which specifies the manner in which the surface changes during plastic flow. Representing this mathematically, we may write

$$F(\sigma_{ij}, \varepsilon_{ij}^{P}, K) = 0 \tag{7.2.6}$$

where the surface is a function of stress, σ_{ij}, plastic strain, ε_{ij}^{P}, and a hardening parameter, K.

Using this definition of the function allows us to conveniently divide the stress state into the following sections:

   (i) an elastic section defined by $F < 0$;
  (ii) a plastic section defined by $F = 0$; and
 (iii) a viscoplastic section defined by $F > 0$.

7.2.3.2   The plastic potential

In viscoplasticity the inelastic strains produced by the state of stress are determined in the same way as in plasticity, i.e. by the use of a flow rule. Classical plasticity theory introduced the concept of a plastic potential [12], which is a scalar function of stress, plastic strains and a hardening parameter, i.e.

$$Q(\sigma_{ij}, \varepsilon_{ij}^p, K) = 0 \qquad (7.2.7)$$

The inelastic strain is determined by partially differentiating the plastic potential with respect to the stresses. To obtain the component of the plastic strain increment in the x-direction, therefore, one would differentiate the plastic potential with respect to the normal stress in the x-direction, i.e.

$$d\varepsilon_{ij}^p = \frac{\partial Q}{\partial \sigma_{ij}} d\lambda \qquad (7.2.8)$$

where $d\lambda$ is a positive constant of proportionality. This defined the direction of plastic straining as being directed along the outward normal to the plastic potential, and was generally known as the normality principle.

Much of classical plasticity theory is based upon the assumption that the material complies with Drucker's definition of a stable inelastic material [13]. Briefly, this requires that the 'work performed by the increments of internal forces on the corresponding increments of the components of the displacement vector must be non-negative'. Convexity of the yield surface and normality of the plastic strain vector to the yield surface are consequences of this initial proposition. Drucker's requirements are satisfied if $F = Q$, and if this is the case the flow rule is said to be associated. The general case can be expressed by defining Q as a separate entity to F which now defines a non-associated flow rule.

7.2.4   Calculation of elasto/viscoplastic strain

The calculation of elasto/viscoplastic strain that follows is basically the same as the general formulation given by Perzyna [14]. The viscoplastic strain rate is assumed to be a unique function of stress, being directly proportional to the excess stress above the yield surface. Therefore, taking into account that the directions of strains are given above by the partial derivative of the plastic potential, the viscoplastic strain rate can be written quite generally as

$$\varepsilon^{vp} = \langle \psi(F) \rangle \frac{dQ}{d\sigma} \qquad (7.2.9)$$

in which

$$\begin{aligned} \langle \psi(F) \rangle &= 0 \qquad \text{if } F \leqslant 0 \\ \langle \psi(F) \rangle &= \psi(F) \quad \text{if } F > 0 \end{aligned} \qquad (7.2.10)$$

Therefore, this equation defines the dependence of the rate of flow on the 'distance' of the stress state from the yield surface. The expression $\psi(F)$ may be written as $\gamma'\Omega(F/F_0)$, where γ' is a fluidity parameter. F_0 denotes any reference value of F, introduced to render the function $\Omega(F/F_0)$ non-dimensional and so to enable the same constant, γ', to be used for any function $\Omega(F/F_0)$.

Perzyna discusses the use of various functions for $\Omega(F/F_0)$ and suggests, amongst others, the following types:

(a) linear law $\Omega(F/F_0) = F/F_0$

(b) power law $\Omega(F/F_0) = (F/F_0)^n$ (7.2.11)

(c) exponential law $\Omega(F/F_0) = (\exp(F/F_0) - 1)$

In the examples quoted here, the power law has been used which reduces to the linear function when $n = 1$.

Combining equations (7.2.3), (7.2.5) and (7.2.9), the complete constitutive relationship for a viscoplastic material may be written as

$$\varepsilon = \mathbf{D}^{-1}\sigma + \langle\psi(F)\rangle\frac{d\mathbf{Q}}{d\sigma} + \varepsilon^i \qquad (7.2.12)$$

7.2.5 The yield criterion used

The yield criterion used here is the Von Mises yield criterion. This can be described as follows. For an isotropic material, plastic yielding can only depend on the magnitudes of the three principal applied stresses and not on their directions. On physical grounds, therefore, any yield criterion should be independent of the co-ordinate system used. Therefore it should be a function of the three stress invariants only:

$$f(J_1, J_2, J_3) = 0 \qquad (7.2.13)$$

where

$$
\begin{aligned}
J_1 &= \sigma_{ii} \\
J_2 &= \tfrac{1}{2}\sigma_{ij}\sigma_{ij} \\
J_3 &= \tfrac{1}{3}\sigma_{ij}\sigma_{jk}\sigma_{ki}
\end{aligned}
\qquad (7.2.14)
$$

Experimental evidence shows that the yielding of a material is, to a first approximation, unaffected by a moderate hydrostatic pressure. Therefore assuming this to be strictly true for an ideal plastic body, it follows that yielding depends only on the deviatoric part of the stresses. Hence the yield criterion must only be a function of J_2' and J_3', since $J_1' = 0$. Therefore the yield criterion reduces to the form

$$f(J_2', J_3') = 0 \qquad (7.2.15)$$

where

$$J'_2 = \tfrac{1}{3}\sigma'_{ij}\sigma'_{ij}$$

$$J'_3 = \tfrac{1}{2}\sigma'_{ij}\sigma'_{jk}\sigma'_{ki}$$

(7.2.16)

where

$$\sigma'_{ij} = \sigma_{ij} - \delta_{ij}(1/3)\sigma_{kk}$$

(7.2.17)

A further restriction is now imposed in view of the assumption that the Bauschinger effect is ignored, so that the magnitude of the yield stress is the same in tension and compression. The yield function can only be an even function of J'_3 since if the signs of the stresses are reversed, the sign of J'_3 is reversed. Hence the most general yielding function is

$$f(J'_2, J'_3) = k'$$

(7.2.18)

where f is an even function of J'_3.

Most of the various yield criteria that have been suggested are no longer of any interest, since they conflict with the requirement that a hydrostatic stress should not influence yielding. However, Von Mises' theory does not have this fault. Von Mises suggested that yielding occurred when J'_2 reached a critical value, or in other words, that the function f in equation (7.2.18) did not involve J'_3. His criterion can be written in the alternative forms:

$$2J'_2 = 2(k')^2$$

(7.2.19)

$$\sigma'^2_1 + \sigma'^2_2 + \sigma'^2_3 = 2(k')^2$$

(7.2.20)

$$(\sigma_1 - \sigma_2)^2 + (\sigma_2 - \sigma_3)^2 + (\sigma_3 - \sigma_1)^2 = 6(k')^2$$

(7.2.21)

$$(\sigma_x - \sigma_y)^2 + (\sigma_y - \sigma_z^2) + (\sigma_z - \sigma_x)^2 + 6(\tau^2_{yz} + \tau^2_{2x} + \tau^2_{xy}) = 6(k)^2$$

(7.2.22)

For the case of uniaxial tension, the yield stress $Y = \sigma_1$ and $\sigma_2 = \sigma_3 = 0$. Substituting into equation (7.2.21) we have

$$2Y^2 = 6(k')^2$$

i.e.

$$k' = Y/\sqrt{3}$$

(7.2.23)

which, when substituted into equation (7.2.19), gives

$$(J'_2)^{1/2} = Y/\sqrt{3}$$

(7.2.24)

i.e.

$$\sqrt{3}(J'_2)^{1/2} - Y = 0$$

(7.2.25)

or, in alternative notation,

$$\sqrt{3}\sigma' - Y = 0$$

(7.2.26)

7.2.6 Finite element solution procedure

The finite element analysis is performed using the same elements and mesh discretisation as used for the heat transfer analysis. If δ denotes the displacement field, then this may be approximated over the domain of interest, Ω, in terms of the nodal displacements, δ_r, by

$$\delta\Omega\delta = \sum_{r=1}^{n} N_r\delta_r = N\delta \tag{7.2.27}$$

where δ is the column vector of nodal values δ_r. With the displacements at all points given by equation (7.2.27), the strains at any point can be determined from the relationship

$$\varepsilon = B\delta \tag{7.2.28}$$

where B is the standard matrix derived from the derivatives of the shape functions [1], and in two dimensions can be written as

$$B = \begin{bmatrix} \dfrac{\partial N_i}{\partial x} & 0 \\[2mm] 0 & \dfrac{\partial N_i}{\partial y} \\[2mm] \dfrac{\partial N_i}{\partial y} & \dfrac{\partial N_i}{\partial x} \end{bmatrix} \tag{7.2.29}$$

With body forces b acting per unit volume and prescribed surface pressure t acting along the surface boundaries, the virtual work principle enables the equilibrium equation to be written as

$$\int_{\Omega} B^T\sigma \, d\Omega - \int_{\Omega} N^T b \, d\Omega - \int_{\Omega} N^T t \, d\Gamma = 0 \tag{7.2.30}$$

with Γ representing the boundary region.

Substituting from (7.2.3):

$$\int_{\Omega} B^T D\varepsilon^e \, d\Omega - \int_{\Omega} N^T b \, d\Omega - \int_{\Gamma} N^T t \, d\Gamma = 0 \tag{7.2.31}$$

i.e.

$$\int_{\Omega} B^T D\varepsilon^e - R_b = 0 \tag{7.2.32}$$

Since the constitutive relationship for viscoplastic problems has been specified in a time rate form, it is convenient to rewrite (7.2.32) as

$$\int_{\Omega} B^T D\dot{\varepsilon}^e - \dot{R}_b = 0 \tag{7.2.33}$$

Substituting for ε^e from (7.2.5) leads to

$$\int_\Omega \mathbf{B}^T\mathbf{D}(\dot{\varepsilon} - \dot{\varepsilon}^{vp} - \dot{\varepsilon}^i)\,d\Omega - \dot{\mathbf{R}}_b - 0 \qquad (7.2.34)$$

which, on using (7.2.28) to substitute for ε, becomes

$$\int_\Omega \mathbf{B}^T\mathbf{D}\mathbf{B}\dot{\delta}\,d\Omega - \int_\Omega \mathbf{B}^T\mathbf{D}\dot{\varepsilon}^{vp}\,d\Omega - \int_\Omega \mathbf{B}^T\mathbf{D}\dot{\varepsilon}^i\,d\Omega - \dot{\mathbf{R}}_b = 0 \qquad (7.2.35)$$

which can be abbreviated to

$$\mathbf{K}\dot{\delta} - \dot{\mathbf{R}} = 0 \qquad (7.2.36)$$

Where $\mathbf{K}$ is the overall stiffness matrix, and $\mathbf{R}$ represents the total loading rate which contains all contributions due to initial strains, viscoplastic strains, body forces and surface pressures.

Equation (7.2.36) is in time varying form and its solution is achieved by representing this time variation as a series of quasi-static solutions. Each of these solutions requires the evaluation of a correction term due to the viscoplastic strains that have occurred since the previous solution. To evaluate this correction term a time stepping procedure is required which evaluates real strains in the material when true viscoplasticity is analysed. For the work described here the viscoplastic algorithm is used to provide elasto-plastic solutions and the time stepping procedure is solely a computational artifice whereby 'steady-state' elasto-plastic values can be obtained.

The time stepping procedure adopted is identical to the initial strain method. The simplest procedure to adopt is the forward difference scheme. By this method the value of a variable at the end of a time step may be predicted from the values at the beginning of the time step. The scheme remains stable provided the time step length does not exceed a critical value.

The essential steps in the solution process can be summarised as follows:

Step 1. The solution to the problem must begin from the known initial conditions at time $t = 0$, which are, of course, the solution of the static elastic problem. At this stage stresses, σ_e^0, displacement, δ^0, load vector, $\mathbf{R}^0$, and viscoplastic strain are known as $(\varepsilon^{vp})^0 = 0$.

Step 2. Calculate the viscoplastic strain rate according to

$$\dot{\varepsilon}^{vp} = \langle \psi(F) \rangle \frac{dQ}{d\sigma} \qquad (7.2.9)$$

Step 3. Calculate the change in viscoplastic strain over a time interval Δt

$$(\Delta\varepsilon^{vp})^t = (\dot{\varepsilon}^{vp})^t\,\Delta t \qquad (7.2.37)$$

Step 4. Add the increment of plastic strain to the total plastic strain at time

t to give the plastic strain at time $t + \Delta t$

$$(\varepsilon^{vp})^{t+\Delta t} = (\varepsilon^{vp})^t + (\Delta \varepsilon^{vp})^t \qquad (7.2.38)$$

Step 5. Calculate the incremental change in R, the loading vector, due to the incremental change $\Delta\varepsilon^{vp}$

$$dR = \int_\Omega B^T D \Delta \varepsilon^{vp} \, d\Omega \qquad (7.2.39)$$

Step 6. Calculate the new loading vector, R, and then calculate the new displacement

$$\delta^{t+\Delta t} = K^{-1}R \qquad (7.2.40)$$

Step 7. From the displacements calculate the total strains

$$\varepsilon^{t+\Delta t} = B\delta^{t+\Delta t} \qquad (7.2.41)$$

Step 8. Calculate the new stresses at time $t + \Delta t$ using the expression

$$\sigma^{t+\Delta t} = D[\varepsilon^{t+\Delta t} - (\varepsilon^{vp})^{t+\Delta t} - \varepsilon^i] + \sigma^i \qquad (7.2.42)$$

Step 9. Return to Step 2 and repeat the process for time level $t + 2\Delta t$.

The forward difference scheme is not unconditionally stable and so the accuracy and stability of the results are directly linked to the size of the time step used. However, theoretical restrictions on the time step length have been provided by Cormeau [15] for specific forms of the viscoplastic flow rule. In particular, for the problem considered here the stability limit is as follows:

$$\Delta t < \frac{4(1 + \nu)F_0}{3\gamma' E} \qquad (7.2.43)$$

where F_0 is the uniaxial yield stress.

It has already been mentioned that the viscoplasticity algorithm is used here for problems of pure plasticity by allowing the solution to progress until stationary conditions are reached. The approach to a steady-state solution is asymptotic, with the viscoplastic strain rate diminishing as the yield surface is approached. To obtain complete convergence would require that, the relationship $F < 0$ is satisfied everywhere within the domain. This condition is seldom reached and so in the solution procedure a tolerance is specified such that when

$$F/F_0 < \text{tolerance} \qquad (7.2.44)$$

the solution is considered to have reached steady-state conditions and the time stepping procedure for that particular load increment is stopped. In practice a tolerance limit of 0.5% of F_0 has been found to be satisfactory.

7.2.7 *Connecting heat transfer theory and stress analysis*

The analysis of heat transfer problems has been described in detail throughout the book, and the solution of the problem provides the distribution of temperature throughout the body at selected intervals of time. Changes in temperature cause thermal expansion and contraction and therefore quite generally we may write

$$\Delta\varepsilon = \alpha_T \Delta T \tag{7.2.45}$$

where $\Delta\varepsilon$ is the change in strain taking place in the body and α_T is the coefficient of thermal expansion. The values of those coefficients are determined from experiments on very small samples and it is therefore important to realise that the strains calculated from their use are unrestrained strains. They do not represent the strains that would take place in a continuum. They do, however, represent initial strains in a body and as such can be converted into stresses and hence forces to be used as the loading vector in the finite element stress analysis. Typically the conversion is made by means of equation (7.2.35) in the elasto/viscoplastic analysis, but the formulation is equally applicable to the linear elastic analysis.

The complete method of solution for the calculation of thermal stresses can therefore be summarised as follows:

Step 1. Calculate the distribution of temperature at times t and $t + \Delta t$, say.

Step 2. Calculate the change in temperature at all nodal points during the time step and convert into initial strains and equivalent nodal 'forces'.

Step 3. Using either an elastic or an elasto/viscoplastic algorithm solve the problem for displacements and strains and calculate the stresses at the end of the time step. These stresses must then be used as initial stresses in the next stress solution.

Step 4. Return to Step 1 and calculate the new distributions of temperature at time $t + 2\Delta t$.

Step 5. Repeat Step 2.

Step 6. Use the stress algorithm again to solve for displacements, strains and stresses at time $t + 2\Delta t$, but at this time step the loading, in terms of equivalent nodal 'forces', consists not only of initial strains but also of initial stresses which are the stresses calculated at the end of the previous time step.

Step 7. Repeat the process at each time step and hence build up a picture of the development of stresses.

It is clear from the above procedure that the stress analysis is 'stress-path' dependent since stresses at the end of one increment are used as initial stresses in the next. This 'stress-path' dependence is not significant in the case of an elastic analysis, since the principle of superimposition holds. However, for a plastic analysis stress-path dependence is of crucial importance. Stresses must

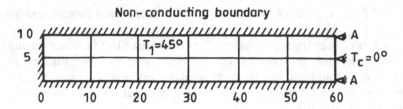

Figure 7.2.3   *Solidification of a bar—problem analysed and finite element mesh.*

be determined at close intervals of change in temperature in order to ensure that the correct stress path is being followed.

7.2.8   Applications

To illustrate the application of these methods three cases of thermal stresses in solidification processes are considered. The first case is a linear elastic analysis of the stress distribution in a solidifying bar [2]. The second case is an elasto/viscoplastic analysis of the continuous casting of metals [3, 16] and the third case is also an elasto/viscoplastic analysis but now of ingot casting [17, 18].

7.2.8.1   Stress distribution in a solidifying bar

The problem analysed is shown diagrammatically in Figure 7.2.3 together with the finite element mesh used. The thermo-physical and mechanical properties used are specified in dimensionless terms as follows:

k = 1.0, thermal conductivity
ρC_p = 2, heat capacity
L = 70.26, ρC_p, volumetric latent heat of fusion
T_i = initial temperature
T_c = 0°, temperature causing solidification on surface AA
E = 1.0, modulus of elasticity
G = 1.38, shear modulus
v = 0.3, Poisson's ratio
α_T = 1.0 × 10^6, coefficient of linear thermal expansion

Results, of the heat transfer analysis are given in Figure 7.2.4 as a plot of the time taken by the cooling front to reach a point $(1 - x)$, versus x, where l is the length of the bar and x is the horizontal distance from the origin. The results are presented in this way to allow comparison to be made with another solution provided by Jones [19]. In general good correlation is obtained between the two sets of results although the rate of cooling predicted by the finite element method is greater the further the cooling front has penetrated into the bar.

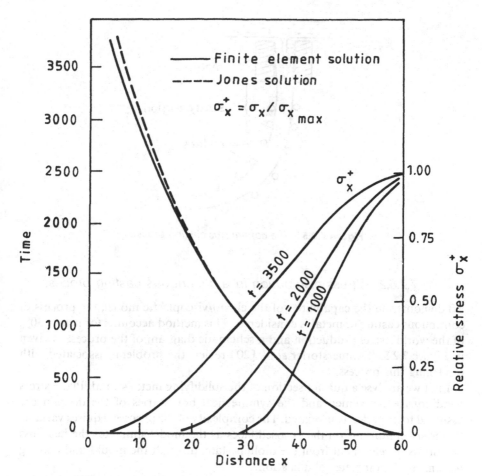

Figure 7.2.4 *The progress of the cooling front and the thermal stress development in the bar.*

To perform a stress analysis of this problem the end AA of the bar is assumed to become solid at $t = 0$ and can therefore be taken as being rigidly fixed. The stresses obtained from the linear elastic analysis are shown in Figure 7.2.4. The distribution of stress along the bar is shown for three different values of time. The stress distribution is plotted in terms of relative stress which is defined as actual normal stress in the x-direction divided by the maximum normal stress in the x-direction. Consequently, the value of the elastic stress is a maximum of unity at face AA and gradually decreases as x diminishes. The pattern of stress distribution obtained reveals clearly the gradual increase in the value of residual stresses that could be expected as the cooling front progresses towards the other end of the bar.

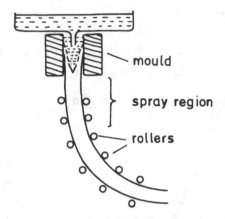

Figure 7.2.5 *The continuous casting process.*

7.2.8.2 *Stress distribution in a continuous casting process*

To demonstrate the capabilities of the elasto/viscoplastic model, the process of continuous casting of metal is considered. This method accounts for nearly 30% of the world's steel production and a schematic diagram of the process is given in Figure 7.2.5. Rammerstorfer *et al.* [20] review the problems associated with modelling this process.

First we analyse a quarter section of the solidifying metal strand. Plane stress conditions are assumed and the symmetrical boundaries of the domain are assumed to be perfectly insulated. The complexity of the process requires variable boundary conditions on the exposed faces as the model simulates the heat loss and stress development from the molten state, through the mould and cooling regions, until complete solidification.

In the mould the heat loss is by conduction only. Within the cooling region two different convective rates are assumed, corresponding to simple air cooling and forced water spraying, respectively. The temperature dependence of various thermal and mechanical properties is given in Figure 7.2.6.

Figure 7.2.7(a) shows the principal stress distribution when the section reaches the end of the mould, both for a purely elastic and an elasto/viscoplastic analysis. Shear stress contours at the same stage in the process are given in Figure 7.2.7(b).

Figure 7.2.8.(a) shows the principal stresses after one-quarter of the air-cooling period and Figure 7.2.8(b) after one-quarter of the water-cooling period.

The numerical model has been used [16] to test an innovation in machine design. Until recently testing was on full-scale machines, which is extremely expensive for the complex casting process outlined above. The thermal stresses shown above often lead to cracking in the metal strand and also wear on the rollers. A new mould design has recently been patented [21] which incorporates a concavity into the faces of the strand. It is intended to accelerate the cooling

THERMAL CONDUCTIVITY

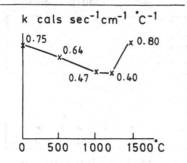

k cals sec^{-1}cm^{-1} $^{\circ}$C^{-1}

0.75

0.64

0.47   0.40

x 0.80

0   500   1000   1500 $^{\circ}$C

THERMAL EXPANSION

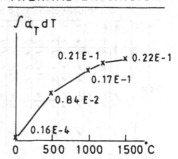

$\int \alpha_T \, dT$

0.21E-1   x 0.22E-1

0.17E-1

0.84 E-2

0.16E-4

0   500   1000   1500 $^{\circ}$C

YOUNG'S MODULUS

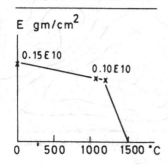

E gm/cm^2

0.15E10

0.10E10

x–x

0   500   1000   1500 $^{\circ}$C

YIELD STRESS

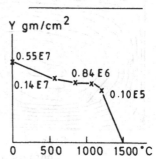

Y gm/cm^2

0.55E7

0.14E7

0.84 E6

0.10E5

0   500   1000   1500 $^{\circ}$C

ENTHALPY

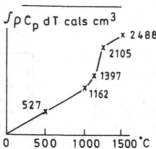

$\int \rho C_p \, dT$ cals cm^3

x 2488

x 2105

x 1397

x 1162

527 x

0   500   1000   1500 $^{\circ}$C

Figure 7.2.6 *The thermal and mechanical properties used in the continuous casting process.*

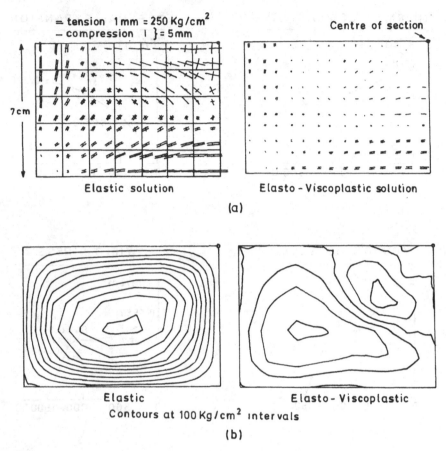

(a)

Elastic solution   Elasto-Viscoplastic solution

Elastic   Elasto-Viscoplastic

Contours at 100 Kg/cm² intervals

(b)

Figure 7.2.7 (a) Principal stresses at the end of the mould. (b) Shear stress contours at the end of the mould.

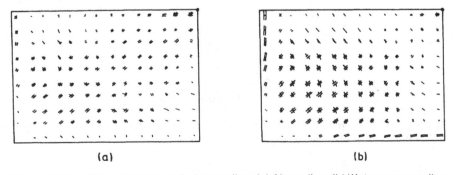

(a)   (b)

Figure 7.2.8 Principal stresses during cooling. (a) Air cooling. (b) Water spray cooling.

and reduce the stresses. It is this design which has been examined by the finite element model. Figure 7.2.9(a) shows the temperature distribution at the end of the mould at $t = 360\,$s for both the straight and concave faced casting process. It is noticeable that the concave strand has cooled much more than the straight strand. The elastic and elasto/viscoplastic stress fields are shown in Figures 7.2.9(b) and (c) for the straight and concave sections, respectively. The stresses in the straight section are almost 15% higher than those in the concave section. These results confirm the expectations of the designer, with the concave section cooling faster yet having lower stresses than the straight section. Thus the concave section is capable of sustaining a faster withdrawal rate, with less

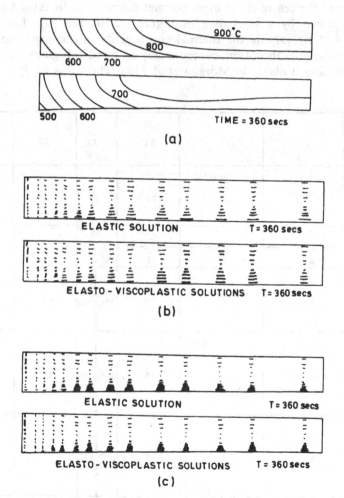

Figure 7.2.9 *(a) Temperature contours at the end of the mould. (b) Stresses at the end of the mould–straight section. (c) Stresses at the end of the mould–concave section.*

chance of crack formation. However, more accurate modelling is required, including the air gap effects in the mould, before further conclusions can be drawn.

7.2.8.3 Stress distribution in the solidification of a steel ingot

The final example considered deals with the solidification of steel ingots. The thermal analysis of this problem can be found in Reference 16. A recent improvement in the modelling of the solidification process is to allow for the formation of an air gap between the ingot and mould. Figure 7.2.10 shows the mesh employed in the thermal stress analysis. The air gap varies in time from the corner to the centre of the ingot perimeter. Since the discretised gap width is fixed (shown by a thick line in Figure 7.2.10), the gap 'formation' is accomplished by varying the thermal conductivity in the gap. Temperature profiles and solidification-front contours may be found in Figures 5.5.10–5.5.12. The details are available in Morgan *et al.* [18]. The stress analysis assumed

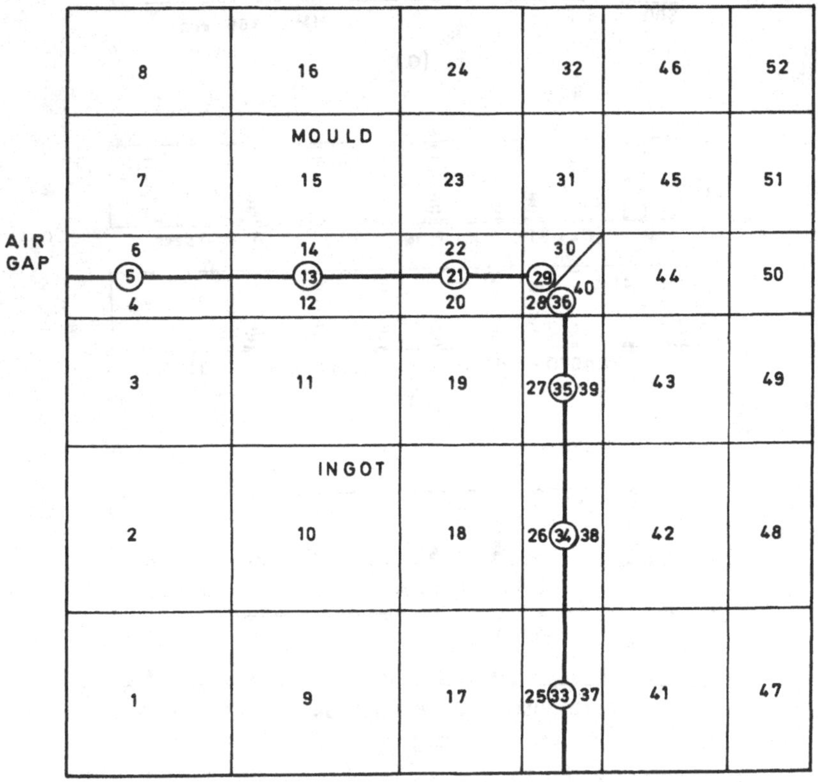

Figure 7.2.10   *Finite element mesh for ingot, air gap and mould.*

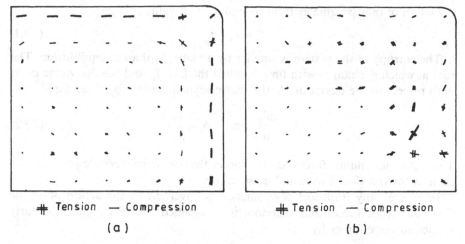

Figure 7.2.11

Figure 7.2.11 *Principal stress field after (a) 4 min; (b) 40 min.*

plane-strain conditions. Figure 7.2.11 depicts the calculated stress distributions in the ingot at two different times. An interesting feature of these results is that the ingot surface is initially subjected to tension but at a later stage this changes to compression.

7.3 HEAT AND MOISTURE TRANSFER

Fundamental considerations leading to the derivation of Luikov's equations for coupled heat and moisture transfer are first presented in this section. Two cases are considered: first, the case of a material's thermophysical parameters varying as functions of the temperature and the moisture content of a material, and second, the case where the material's thermophysical parameters may be assumed to be constant in the range of interest. Details of the finite element solution of both cases are then presented before proceeding to the application of the model to practical engineering problems.

7.3.1 Potentials for transfer and transfer fluxes

To extend heat transfer to encompass coupled analysis of heat and moisture transfer, two universal laws of nature, namely the law of conservation and transformation of energy and the law of conservation of mass, are considered. To describe in general terms the two processes of transfer, the concept of entropy, as discussed by Onsager [22], is also introduced. Now for any adiabatically isolated system we can define the state of that system by a parameter π_i, where π_i may be pressure, temperature or concentration, for example. Let θ_i be the deviation of that parameter from its equilibrium value π_i^*:

$$\theta_i = \pi_i - \pi_i^* \tag{7.3.1}$$

The entropy of the system is smaller by a value Δs than at equilibrium. The rate at which θ_i changes with time is called the flux, j_i, and the derivative of Δs with respect to the deviation θ_i, the thermodynamic force ϕ_i. Therefore

$$j_i = \frac{d\theta_i}{dt} \quad \text{and} \quad \phi_i = \frac{\partial s}{\partial \theta_i} \tag{7.3.2}$$

These thermodynamic forces are the forces that cause irreversible phenomena, such as the transfer of heat and mass, to occur.

In general, any transport phenomenon is governed by the action of all the thermodynamic forces and therefore it is assumed that each flux is linearly related to these forces by

$$j_i = \sum_{e=1}^{n} L_{ie}\phi_e \tag{7.3.3}$$

This expression is known as Onsager's system of linear equations and it is the principal expression of the thermodynamics of irreversible processes. The coefficients L_{ie} are called 'phenomenological' because they are determined by the rate at which the phenomena proceed. These coefficients are symmetrical, and thus the cross-coefficients are equal:

$$L_{ie} = L_{ei} \tag{7.3.4}$$

Therefore for the particular case under consideration here, namely coupled heat and moisture transfer, the equations can be written as

$$j_q = L_{11}\phi_1 + L_{12}\phi_2 \quad \text{(heat transfer)}$$

$$j_m = L_{21}\phi_1 + L_{22}\phi_2 \quad \text{(mass transfer)} \tag{7.3.5}$$

The exact form of the thermodynamic forces to be used in Onsager's expression, however, still has to be determined. Onsager's theorem is based on the principle of increase of entropy. From (7.3.2) the rate of increase of entropy with time becomes

$$\frac{ds}{dt} = \sum_{i=1}^{n} j_i\phi_i \tag{7.3.6}$$

i.e. the sum of the products of the fluxes and the corresponding thermodynamic forces. Therefore this equation is the basis for the choice of the thermodynamic forces to be used.

For example, for heat transfer without mass exchange

$$T\,ds = dU'$$

where T is the temperature and U' is the internal energy. Therefore

$$\phi_T = \nabla \left(\frac{\partial s}{\partial U'} \right)_m = D(1/T) = -(1/T)^2 DT \tag{7.3.7}$$

and for isothermal self-diffusion

$$T \, ds = \mu \, dm$$

where μ is the chemical potential and m the mass content

$$\phi_m = \nabla \left(\frac{\partial s}{\partial m} \right)_\mu = -\nabla(\mu/T) \tag{7.3.8}$$

Therefore the thermodynamic force of heat transfer is directly proportional to the temperature gradient, but the mass transfer force depends upon the chemical potential and the temperature.

Substituting the thermodynamic forces given by (7.3.7) and (7.3.8) into (7.3.5) yields

$$j_q = \frac{-L_{11}}{T^2} \nabla T - L_{12} \nabla(\mu/T)$$

$$j_m = \frac{-L_{21}}{T^2} \nabla T - L_{22} \nabla(\mu/T) \tag{7.3.9}$$

Equation (7.3.9) states that heat transfer depends not only on thermal conduction but also on the redistribution of mass (Dufour effect) while mass transfer is governed not only by differences in chemical potential, and thus concentration of matter, but also by thermal diffusion (Soret effect). The cross-coefficients L_{21} and L_{12} are equal; the effect of unequal concentrations of matter on heat-energy flow is symmetrical to that of temperature differences on mass flow. These are therefore the governing expressions for heat and mass flux using the correctly defined thermodynamic forces.

Unfortunately these equations are not easily applied since the chemical potential is not an easily determined quantity. More useful expressions have been obtained by Luikov by replacing the thermodynamic force, μ/T, by moisture content. This approach yields

$$j_q = -k_{11} \nabla T - k_{12} \nabla m$$

$$j_m = -k_{21} \nabla T - k_{22} \nabla m \tag{7.3.10}$$

The phenomenological coefficients, L_{il}, have now been replaced by conductivities, k_{il}. It should be noted that k_{12} is not equal to k_{21} and the symmetry of the equations has been lost.

The Dufour effect usually is considered to be of little significance in drying practice. Therefore ignoring this and rewriting the equations in the form proposed by Luikov produces

$$j_q = -k\nabla T$$

(7.3.11)

$$j_m - a_m\rho\delta^*\nabla T - a_m\rho\nabla m$$

where k is the coefficient of thermal conductivity, ρ is the density and a_m is the mass transfer coefficient. δ^*, the thermogradient coefficient, is a measure of the rate at which moisture moves because of the influence of the temperature gradient.

7.3.2 Conservation equations

In order to define diffusion and filtration moisture transfer, three interdependent partial differential equations with pressure, temperature and moisture content as the working variables will be necessary. In the derivation of these equations the following assumptions are made:

- The mass is present only as liquid and vapour.
- The movement of moisture in the capillary porous body is sufficiently slow so that in practice the temperatures of the liquid, the vapour and the body are equal at coincident points.
- Chemical reactions associated with water loss are not taken into account.
- Dimensional changes that occur within the material, due to a temperature or moisture content change, are comparatively small and can be ignored.

The following subscripts are used to describe the various components of the material.

0—porous body skeleton,
1—vapour,
2—liquid,
3—solid,
4—inert gas.

7.3.2.1 Mass transfer equation

The mass balance for one of the bound materials, vapour or liquid, in the capillary porous bodies, follows from the law of mass conservation as described in the work by Luikov [9]:

$$\frac{\partial(m_i\rho_0)}{\partial t} = -\operatorname{div}(j_{i\mathrm{dif}} + j_{1\mathrm{fil}}) + I_i$$

(7.3.12)

The mass flux is not only related to the gradient of moisture concentration, but also to the temperature gradient. This effect is more commonly known as the

Soret effect, which is described by

$$j_{\text{dif}} = -\alpha_{m\rho_0}(\nabla m + \delta \nabla T) \tag{7.3.13}$$

The presence of a total pressure gradient within the capillary–porous body brings about an additional form of mass transfer of the vapour and liquid mixture, namely by filtration. This filtration transfer is described as follows:

$$j_{\text{fil}} = j_{1\text{fil}} + j_{2\text{fil}} = -k_p \nabla P \tag{7.3.14}$$

Equation (7.3.12) represents the mass balance for the ith bound material. However, the total mass balance for the material is obtained by summing for all the bound materials, $i = 0, 1, 2, 3, 4$:

$$\frac{\partial(\sum m_i \rho_0)}{\partial t} = -\text{div}\left(\sum_{i=0}^{4} j_{i\text{dif}} + \sum_{i=0}^{4} j_{i\text{fil}}\right) + \sum_{i=0}^{4} I_i \tag{7.3.15}$$

However, if we assume that the moisture within the capillary–porous body only exists as vapour and liquid, then we only need to consider the conservation of mass for these two bound materials. The sum of the sources and sinks equals zero, since moisture is neither lost nor gained by the material:

$$\sum_{i=0}^{2} I_i = 0 \tag{7.3.16}$$

Therefore, the total mass balance for the material can be described by rewriting equation (7.3.15) as

$$\frac{\partial(m\rho_0)}{\partial t} = -\text{div}(j_{\text{dif}} + j_{\text{fil}}) \tag{7.3.17}$$

Substituting equations (7.3.13) and (7.3.14) into equation (7.3.17) gives

$$\frac{\partial(m\rho_0)}{\partial t} = \text{div}(\alpha_m \rho_0 \nabla m + \alpha_m \rho_0 \delta \nabla i + k_p \nabla P) \tag{7.3.18}$$

The dry density of the material remains constant and does not vary with time; hence, equation (7.3.18) can be expressed as

$$\frac{\partial m}{\partial t} = \alpha_m \nabla^2 m + \alpha_m \delta \nabla^2 T + \frac{k_p}{\rho_0} \nabla^2 P \tag{7.3.19}$$

7.3.2.2 Heat transfer equation

The balance of the thermal energy within the capillary-porous body is represented by

$$\rho_0 C_q \frac{\partial T}{\partial t} = -\text{div} j_q - \sum_{i=0}^{2} h_i I_i \tag{7.3.20}$$

The transfer of heat by conduction is normally related to the temperature gradient, but in the case of a coupled system it is also related, albeit weakly, to the moisture potential gradient, and this is more commonly known as the Dufour effect. However, usually this is considered to be insignificant for capillary-porous bodies. We can, therefore, write down the transfer of heat by conduction, known as Fourier's law, as

$$j_q = -k_q \nabla T \tag{7.3.21}$$

For the range of moisture contents considered, the mass of the vapour contained within the capillaries of the body is negligible in comparison with the mass of the contained liquid:

$$m = m_1 + m_2 = m_2 \tag{7.3.22}$$

Therefore, from equation (7.3.12):

$$\frac{\partial(\rho_0 m_1)}{\partial t} = 0 = -\operatorname{div} j_1 + I_i \tag{7.3.23}$$

Hence

$$-I_i = -\operatorname{div} j_1 = I_2 \tag{7.3.24}$$

$$\frac{\partial(\rho_0 m_2)}{\partial t} = -\operatorname{div} j_2 + I_2 \tag{7.3.25}$$

When the source of vapour is dependent upon the vaporisation of water, a total change in moisture content, dm, occurs due to the sum of the change in moisture transfer, dm_2, and the change in liquid vaporisation, dm_1:

$$dm = dm_1 + dm_2 \tag{7.3.26}$$

Introducing the factors β and ε according to the following relationships:

$$\beta = \frac{dm_1}{dm_2} \quad \text{and} \quad \varepsilon = \frac{\beta}{1 + \beta} \tag{7.3.27}$$

we obtain

$$\frac{dm_1}{dm_2} = \frac{\varepsilon}{1 + \varepsilon} \tag{7.3.28}$$

By substituting equation (7.3.28) into equation (7.3.26) we obtain a relationship for the vapour diffusion coefficient:

$$\frac{dm_1}{dm} = \varepsilon \tag{7.3.29}$$

In the case of non-steady-state moisture transfer ($dm = 0$), the ratio described by equation (7.3.29) is finite and describes the change in moisture content due to evaporation relative to the total change in moisture content at a given point in the body. If $\varepsilon = 1$ ($\beta = \infty$), then the transfer of moisture occurs in the form of vapour. Since the moisture content of the body, m, is equal to the moisture content of the liquid ($m = m_2$), then a change in the moisture content of the body occurs due to evaporation of liquid or condensation of vapour; transfer of liquid is absent. If $\varepsilon = 0$ ($\beta = 0$), then the moisture content changes only as a result of liquid transfer. Therefore, differentiating equation (7.3.28) with respect to τ and substituting into equation (7.3.26):

$$\frac{\partial(\rho_0 m_2)}{\partial t} = \frac{\partial_1(\rho_0 m_2)}{\partial t} + \frac{\partial_2(\rho_0 m_2)}{\partial t} \tag{7.3.30}$$

$$= \frac{\varepsilon}{1-\varepsilon}\frac{\partial_2(\rho_0 m_2)}{\partial t} + \frac{\partial_2(\rho_0 m_2)}{\partial t}$$

$$= \frac{\varepsilon}{1-\varepsilon}\frac{\partial_2(\rho_0 m_2)}{\partial t}$$

$$= \frac{\varepsilon}{1-\varepsilon}\operatorname{div} j_2$$

Rearranging, we get

$$\frac{\partial_2(\rho_0 m_2)}{\partial t} = -\operatorname{div} j_2 + \varepsilon\frac{\partial(\rho_0 m_2)}{\partial t} . \tag{7.3.31}$$

By comparing equation (7.3.31) with equation (7.3.32) we see that

$$I_2 = \varepsilon\rho_0\frac{\partial m_2}{\partial t} \tag{7.3.32}$$

Substituting equations (7.1.21) and (7.1.24) into the expression for the conservation of thermal energy, equation (7.3.20), we obtain

$$\rho_0 C_q\frac{\partial T}{\partial t} = \operatorname{div}(k_q\nabla T) + I_2(h_1 - h_2) \tag{7.3.33}$$

The difference in the specific enthalpies of the moisture in the designated state, $h_1 - h_2$, is equal to the latent heat of vaporisation, λ. By introducing equation (7.3.32) into equation (7.3.33) we see that

$$\rho_0 C_q\frac{\partial T}{\partial t} = \operatorname{div}(k_q\nabla T) + \varepsilon\lambda\rho_0\frac{\partial m_2}{\partial t} \tag{7.3.34}$$

As previously stated, the mass of the contained vapour is negligible compared with the mass of the liquid, $m = m_2$, and substituting in equation (7.3.19) we obtain the coupled differential equation for heat transfer:

$$\rho_0 C_q\frac{\partial T}{\partial t} = \operatorname{div}(k_q\nabla T) + \varepsilon\lambda\rho_0\left(\alpha_m\delta\nabla^2 T + \alpha_m\nabla^2 m + \frac{k_p}{\rho_0}\nabla^2 P\right) \tag{7.3.35}$$

Rearranging:

$$\rho_0 C_q \frac{\partial T}{\partial t} = (k_q + \varepsilon\lambda\alpha_m\delta\rho_0)\nabla^2 T + \varepsilon\lambda\alpha_m\rho_0\nabla^2 m + \varepsilon\lambda k_p\nabla^2 P \qquad (7.3.36)$$

7.3.2.3 Pressure equation

For a closed system, an equation representing the change in pressure is required. diffusion of vapour and air in the capillaries is small in comparison with filtration transfer. The transfer of moist air within the capillaries is described by

$$j_1 + j_4 = -k_p\nabla P \qquad (7.3.37)$$

Summing the differential equations of mass transfer with respect to $i = 1$ and 4, we obtain

$$\frac{\partial(\rho_0(m_1 + m_4))}{\partial t} = -\text{div}\,(j_1 + j_4) + I_1 + I_4 \qquad (7.3.38)$$

If no chemical changes connected with the formation of a condensing gas (dry air) occur, the source I_4 will be absent:

$$I_4 = 0 \qquad (7.3.39)$$

Introducing equations (7.3.24) and (7.3.39) into equation (7.3.38) we obtain

$$\frac{\partial(\rho_0(m_1 + m_4))}{\partial t} = -\text{div}\,(j_1 + j_4) + I_1 \qquad (7.3.40)$$

The volumetric concentration, ω, of the bound mass (inert gas, vapour, liquid and ice) is equal to the ratio of the mass, w, of the substance to the volume of the body, V:

$$\omega = \frac{w}{V} = \frac{1}{V}\sum_{i=1}^{4} w_i = \sum_{i=0}^{4} \omega_i \qquad (7.3.41)$$

Instead of the volumetric concentration of the bound substance, the relative concentration or the specific mass content, m, is used, where m is the ratio of the mass of the bound substance to the mass of the dry body, w_0 (mass of the solid part of the body). The specific mass content is equal to the sum of the specific mass contents of the bound substances in the different states:

$$m = \sum_{i=1}^{4} m_i = \frac{1}{w_0}\sum_{i=1}^{4} w_i = \frac{V}{w_0}\sum_{i=1}^{4} \omega_i = \frac{1}{\rho_0}\sum_{i=1}^{4} \frac{\omega_i}{\rho_i} \qquad (7.3.42)$$

Therefore, for a vapour–gas mixture the specific mass content can be represented by

$$\rho_0(m_1 + m_4) = \omega_1 + \omega_4 = \rho_g\Pi b \qquad (7.3.43)$$

where ρ_g is the density of the gas–vapour mixture. However, for an ideal gas the density can be expressed as follows:

$$\rho_g = \frac{PM}{RT} \tag{7.3.44}$$

The specific mass content of the vapour–gas mixture is determined by the pressure and temperature:

$$\rho_0(m_1 + m_4) = \frac{PM}{RT} \Pi b \tag{7.3.45}$$

Differentiating equation (7.3.45) we obtain

$$\frac{\partial(\rho_0(m_1 + m_4))}{\partial t} = \frac{M\Pi}{R}\left(\frac{P}{T}\frac{db}{dt} + \frac{b}{T}\frac{dP}{dt} - \frac{Pb}{T^2}\frac{dT}{dt}\right) \tag{7.3.46}$$

However, $T^2 \gg b$ and $T \gg db$, therefore equation (7.3.45) can be simplified by letting

$$C_p = \frac{M\Pi b}{RT} \tag{7.3.47}$$

become

$$\frac{\partial(m_1 + m_4)}{\partial t} = C_p \frac{\partial P}{\partial t} \tag{7.3.48}$$

Substituting equations (7.3.32), (7.3.37) and (7.3.48) into equation (7.3.40):

$$\rho_0 C_p \frac{\partial P}{\partial t} = \operatorname{div}(k_p \nabla P) - \varepsilon\rho_0 \frac{\partial m}{\partial t} \tag{7.3.49}$$

and introducing the partial differential equation for mass transfer into equation (7.3.49):

$$\rho_0 C_p \frac{\partial P}{\partial t} = \operatorname{div}(k_p \nabla P) - \varepsilon\rho_0\left(\alpha_m \nabla^2 m + \alpha_m \delta \nabla^2 T + \frac{k_p}{\rho_0}\nabla^2 P\right) \tag{7.3.50}$$

Rearranging:

$$\rho_0 C_p \frac{\partial P}{\partial t} = -\varepsilon\rho_0\alpha_m\delta\nabla^2 T - \varepsilon\rho_0\alpha_m\nabla^2 m + k_p(1-\varepsilon)\nabla^2 P \tag{7.3.51}$$

Equations (7.3.19), (7.3.36) and (7.3.51) are the governing differential equations for heat and mass transfer under the influence of a total pressure gradient in a capillary-porous body. In these equations the coefficients all vary as functions of temperature, moisture content and pressure in the body.

The coupling coefficients in the above equations are different and consequently when a finite element solution of the equations is assembled non-symmetric matrices are produced. This results in an increase in the required amount of computer time compared with the solution of symmetric problems. If this feature is undesirable, progress can be made by assuming that some of the material properties can be taken to be constant in the range of interest. This approach allows a symmetric solution to be achieved and this method will also be described here.

In this approach the concept of moisture potential is used, defined by Luikov as follows. For constant specific mass capacity, moisture content and moisture potential are related by

$$m = c_m u \tag{7.3.52}$$

The importance of the use of a moisture potential lies in the fact that heat and mass transfer in capillary–porous bodies is determined by a difference of transfer potentials. The heat transfer potential, temperature, is universally established and in order to make the theory of mass transfer compatible, a mass transfer potential is introduced. Its importance is demonstrated in the analysis of drying of laminated materials composed of inhomogeneous moist bodies in contact. When two such bodies are in thermal equilibrium the temperature in both is the same but the heat content is very different and undergoes a rapid change on the boundary of contact. Similarly, in the case of hygrothermal equilibrium the moisture content of the two bodies in contact may be different but the transfer potential must be the same. Consequently, the use of a moisture potential is required. The concept of specific mass capacity is introduced to complete the similarity to the heat transfer problem, specific mass capacity being analogous to specific heat capacity. It follows from the above that the density of flow of moisture is directly proportional to the gradient of the mass transfer potential, i.e.

$$j_m = - k_m \nabla u \tag{7.3.53}$$

where k_m is called the coefficient of mass conductivity.

It can be shown that

$$a_m \rho c_m = k_m \tag{7.3.54}$$

and therefore equations (7.3.19), (7.3.36) and (7.3.51) may be written as

$$\rho_0 c_q \frac{\partial T}{\partial t} = (k_q + \varepsilon' \lambda k_m \delta') \nabla^2 T + \varepsilon' \lambda k_m \nabla^2 u + \varepsilon \nabla k_p \nabla^2 p \tag{7.3.55a}$$

$$\rho_0 c_m \frac{\partial T}{\partial t} = (k_m \delta') \nabla^2 T + k_m \nabla^2 u + k_p \nabla^2 p \tag{7.3.55b}$$

$$\rho_0 c_p \frac{\partial p}{\partial t} = \varepsilon' k_m \delta' \nabla^2 T - \varepsilon' k_m \nabla^2 u + k_p (1 - \varepsilon') \nabla^2 p \tag{7.3.55c}$$

If equation (7.3.55a) is multiplied by δ', equation (7.3.55b) by $\varepsilon \lambda$ and equation

(7.3.55c) by $-\lambda k_p / k_m$, a symmetric set of equations is established as follows:

$$\rho_0 c_q \delta' \frac{\partial T}{\partial t} = (k_q + \varepsilon \lambda k_m \delta') \delta' \nabla^2 T + \varepsilon \lambda k_m \delta' \nabla^2 u + \varepsilon \lambda k_p \delta' \nabla^2 p \qquad (7.3.56\text{a})$$

$$\varepsilon \lambda \rho_0 c_m \frac{\partial u}{\partial t} = \varepsilon \lambda k_m \delta' \nabla^2 T + \varepsilon \lambda k_m \nabla^2 u + \varepsilon \lambda k_p \nabla^2 p \qquad (7.3.56\text{b})$$

$$-\frac{\lambda k_p \rho_0 c_p}{k_m} \frac{\partial p}{\partial t} = \varepsilon \lambda k_p \delta' \nabla^2 T + \varepsilon \lambda k_p \nabla^2 u + \frac{\lambda k_p^2}{k_m} (\varepsilon - 1) \nabla^2 p \qquad (7.3.56\text{c})$$

These equations are the governing differential equations for the fully linear case where all the thermophysical parameters are constant. The set of equations (7.3.56a), (7.3.56b) and (7.3.56c) can now be rewritten conveniently as

$$c_q \frac{\partial T}{\partial t} = \nabla [k_{11} \nabla T + k_{12} \nabla (m \text{ or } u) + k_{13} \nabla p] \qquad (7.3.57\text{a})$$

$$c_m \frac{\partial (m \text{ or } u)}{\partial t} = \nabla [K_{21} \nabla T + K_{22} \nabla (m \text{ or } u) + K_{23} \nabla p] \qquad (7.3.57\text{b})$$

$$c_p \frac{\partial p}{\partial t} = \nabla [K_{31} \nabla T + K_{32} \nabla (m \text{ or } u) + K_{33} \nabla p] \qquad (7.3.57\text{c})$$

7.3.3 Boundary conditions

A general set of boundary conditions for the system of equations is given by

$$T = T_w \quad \text{on } \Gamma_1 \qquad (7.3.58)$$

on the portion of the boundary with a constant temperature, and

$$k \frac{\partial T}{\partial n} + j_q + \alpha_q (T - T_a) + (1 - \varepsilon') \lambda \alpha_m \rho (m - m_a) = 0 \quad \text{on } \Gamma_2 \qquad (7.3.59)$$

the part of the boundary subjected to heat flux conditions where T_a and m_a are ambient temperature and moisture content, respectively, α_q is the convective heat transfer coefficient, α_m the convective mass transfer coefficient in terms of moisture content differences, and Γ_1 and Γ_2 make up the complete boundary surface. The coefficients of convective heat and mass transfer are assumed to vary with respect to their primary variables.

The first term of equation (7.3.59) corresponds to the amount of heat passing into the body, the second and third terms to the amount of heat supplied at the surface of the body, and the last term is the amount of heat expended in the evaporation of liquid. The second and third terms can be subdivided into the case where the flow of heat is constant and into the case where the flow of heat is dependent on the difference between the body and ambient temperatures.

For the mass transfer, we have

$$m = m_w \quad \text{on } \Gamma_3 \tag{7.3.60}$$

the portion of the boundary with a constant moisture content and

$$a_m\rho \frac{\partial m}{\partial n} + a_m\rho\delta^* \frac{\partial T}{\partial n} + j_m + \alpha_m\rho(m - m_a) = 0 \quad \text{on } \Gamma_4 \tag{7.3.61}$$

the part of the boundary subjected to a specified moisture flux. Equation (7.3.61) reflects the mass balance at the surface, where the first two terms describe the supply of moisture to the surface under the gradients of temperature and moisture content and the final two terms describe the way in which the moisture is drawn off from the surface.

For the linear case, the boundary conditions are slightly modified to take account of the use of moisture potential and are given as follows:

$$T = T_w \quad \text{on } \Gamma_1 \tag{7.3.62}$$

$$k\frac{\partial T}{\partial n} + j_q + \alpha_q(T - T_a) + (1 - \varepsilon')\lambda\alpha_u(u - u_a) = 0 \quad \text{on } \Gamma_2 \tag{7.3.63}$$

$$U = U_w \quad \text{on } \Gamma_3 \tag{7.3.64}$$

$$k\frac{\partial u}{\partial n} + j_m + k_m\delta' \frac{\partial T}{\partial n} + \alpha_u(u - u_a) = 0 \quad \text{on } \Gamma_4 \tag{7.3.65}$$

$$P = P_w \quad \text{on } \Gamma_5 \tag{7.3.66}$$

the portion of the boundary with a constant total gas pressure where α_u is the convective mass transfer coefficient in terms of moisture potential and is related to α_m by

$$\alpha_m = \alpha_u/\rho c_m \tag{7.3.67}$$

Both of these sets of equations can be written in the same two-dimensional generalised form. In particular, the flux boundary conditions can be rearranged to produce

$$K_q \frac{\partial T}{\partial n} + J_q^* = 0 \tag{7.3.68}$$

$$K_m \frac{\partial(m \text{ or } u)}{\partial n} + J_m^* = 0 \tag{7.3.69}$$

A full definition of the form of the generalised fluxes J_q^* and J_m^* can be found in Thomas [24].

Since both sets of governing equations have been written in the same form the finite element formulation that follows can be applied to either case. For the sake of convenience, the formulation will be presented in terms of u, the moisture potential, but it is equally applicable to moisture content, m.

7.3.4 *Finite element formulation*

The finite element formulation presented here can be applied to both the fully non-linear and the fully linear systems of coupled partial differential equations since the systems of equations have been presented in the same form. The formulation is described in terms of moisture content, i.e. the fully non-linear formulation.

The variation of the temperature, pressure and moisture content throughout the domain of interest, Ω, is approximated in terms of the nodal values T_s, m_s and P_s as

$$T \approx \sum_{s=1}^{n} N_s(x, y) T_s(t)$$

$$m \approx \sum_{s=1}^{n} N_s(x, y) m_s(t) \qquad (7.3.70)$$

$$P \approx \sum_{s=1}^{n} N_s(x, y) P_s(t)$$

If the approximations given by equations (7.3.70) are substituted into equations (7.3.57), a residual is obtained, which is then minimised using the Galerkin method. This requires that the integral of the weighted errors over the domain Ω, must be zero, with the shape functions N_r being used as the weighting functions, i.e.

$$\int_{\Omega} N_r \left[\nabla \cdot (K_{11} \nabla T) + \nabla \cdot (K_{12} \nabla m) + \nabla \cdot (K_{13} \nabla P) - c_q \frac{\partial T}{\partial t} \right] d\Omega = 0 \quad (7.3.71)$$

$$\int_{\Omega} N_r \left[\nabla \cdot (K_{21} \nabla T) + \nabla \cdot (K_{22} \nabla m) + \nabla \cdot (K_{23} \nabla P) - c_m \frac{\partial m}{\partial t} \right] d\Omega = 0 \quad (7.3.72)$$

$$\int_{\Omega} N_r \left[\nabla \cdot (K_{31} \nabla T) + \nabla \cdot (K_{32} \nabla m) + \nabla \cdot (K_{33} \nabla P) - c_p \frac{\partial P}{\partial t} \right] d\Omega = 0 \quad (7.3.73)$$

The application of Green's theorem (integration by parts) to reduce the order of the above equations, to equation (7.3.71), and expanding gives

$$-\int_{\Omega} \left\{ K_{11} \left[\frac{\partial N_r}{\partial x} \frac{\partial T}{\partial x} + \frac{\partial N_r}{\partial y} \frac{\partial T}{\partial y} \right] + K_{12} \left[\frac{\partial N_r}{\partial x} \frac{\partial m}{\partial x} + \frac{\partial N_r}{\partial y} \frac{\partial m}{\partial y} \right] \right.$$

$$\left. + K_{13} \left[\frac{\partial N_r}{\partial x} \frac{\partial P}{\partial x} + \frac{\partial N_r}{\partial y} \frac{\partial P}{\partial y} \right] - c_q N_r \frac{\partial T}{\partial t} \right\} d\Omega + \int_{\Gamma} \left\{ K_{11} N_r \left[n_x \frac{\partial T}{\partial x} + n_y \frac{\partial T}{\partial y} \right] \right.$$

$$\left. + K_{12} N_r \left[n_x \frac{\partial m}{\partial x} + n_y \frac{\partial m}{\partial y} \right] + K_{13} N_r \left[n_x \frac{\partial P}{\partial x} + n_y \frac{\partial P}{\partial y} \right] \right\} d\Gamma = 0 \qquad (7.3.74)$$

Having reduced the derivatives from second order to first order, we can substitute into equation (7.3.74) the nodal values T_s, m_s and P_s from equation (7.3.70) and the generalised boundary conditions from equations (7.3.41) and (7.3.42):

$$\int_\Omega [K_{11}T_s + K_{12}m_s + K_{13}P_s]\left[\frac{\partial N_r}{\partial x}\frac{\partial N_s}{\partial x} + \frac{\partial N_r}{\partial y}\frac{\partial N_s}{\partial y}\right]$$

$$+ c_q N_r N_s \frac{\partial T_s}{\partial t}d\Omega + \int_\Gamma N_r\left[J_q^* + \frac{K_{12}}{K_{22}}J_m^*\right]d\Gamma = 0 \qquad (7.3.75)$$

Equations (7.3.72) and (7.3.73) can be expressed in a similar manner:

$$\int_\Omega [K_{21}T_s + K_{22}m_s + K_{23}P_s]\left[\frac{\partial N_r}{\partial x}\frac{\partial N_s}{\partial x} + \frac{\partial N_r}{\partial y}\frac{\partial N_s}{\partial y}\right]$$

$$+ c_m N_r N_s \frac{\partial m_s}{\partial t}d\Omega + \int_\Gamma N_r\left[\frac{K_{21}}{K_{11}}J_q^* + J_m^*\right]d\Gamma = 0 \qquad (7.3.76)$$

$$\int_\Omega [K_{31}T_s + K_{32}m_s + K_{33}P_s]\left[\frac{\partial N_r}{\partial x}\frac{\partial N_s}{\partial x} + \frac{\partial N_r}{\partial y}\frac{\partial N_s}{\partial y}\right]$$

$$+ c_p N_r N_s \frac{\partial P_s}{\partial t}d\Omega + \int_\Gamma N_r\left[\frac{K_{31}}{K_{11}}J_q^* + \frac{K_{32}}{K_{22}}J_m^*\right]d\Gamma = 0 \qquad (7.3.77)$$

Equations (7.3.75), (7.3.76) and (7.3.77) may be rewritten in the following matrix form:

$$K(\Phi)\Phi + C(\Phi)\frac{\partial \Phi}{\partial t} + J(\Phi) = 0 \qquad (7.3.78)$$

where $K(\Phi)$ and $C(\Phi)$ are solution-dependent matrices:

$$\mathbf{K} = \begin{bmatrix} K_{11} & K_{12} & K_{13} \\ K_{21} & K_{22} & K_{23} \\ K_{31} & K_{32} & K_{33} \end{bmatrix}$$

$$\mathbf{C} = \begin{bmatrix} C_q & 0 & 0 \\ 0 & C_m & 0 \\ 0 & 0 & C_p \end{bmatrix}$$

$$\mathbf{\Phi} = \begin{bmatrix} T \\ m \\ P \end{bmatrix}$$

$$\mathbf{J} = \begin{bmatrix} J_q \\ J_m \\ J_P \end{bmatrix}$$

Typical matrix elements are

$$K_{rs} = \sum_{r,s=1}^{n} \int_{\Omega^e} K\left(\frac{\partial N_r}{\partial x}\frac{\partial N_s}{\partial x} + \frac{\partial N_r}{\partial y}\frac{\partial N_s}{\partial y}\right) d\Omega$$

$$C_{rs} = \sum_{r,s=1}^{n} \int_{\Omega^e} CN_r N_s \, d\Omega$$

$$J_q = \sum_{r=1}^{n} \int_{\Gamma^e} N_r\left(J_q^* + \frac{K_{12}}{K_{22}}J_m^*\right) d\Gamma$$

$$J_m = \sum_{r=1}^{n} \int_{\Gamma^e} N_r\left(\frac{K_{21}}{K_{11}}J_q^* + J_m^*\right) d\Gamma$$

$$J_p = \sum_{r=1}^{n} \int_{\Gamma^e} N_r\left(\frac{K_{31}}{K_{11}}J_q^* + \frac{K_{32}}{K_{22}}J_m^*\right) d\Gamma$$

Equation (7.3.78) represents the spatially discretised form of the governing equation. To calculate the time variation of ϕ, equation (7.3.78) is replaced by a finite difference recurrence relationship. In particular, Lees' three-level time stepping scheme has been applied for this work, leading to the relationship

$$K^n\frac{[\phi^{n+1} + \phi^n + \phi^{n-1}]}{3} + C^n\frac{[\phi^{n+1} - \phi^{n-1}]}{2\Delta t} + J^n = 0$$

which, on rearranging, produces the recurrence relationship

$$\phi^{n+1} = -\left[\frac{K^n}{3} + \frac{C^n}{2\Delta t}\right]^{-1}\left[\frac{K^n\phi^n}{3} + \frac{K^n\phi^{n-1}}{3} - \frac{C^n\phi^{n-1}}{2\Delta t} + J^n\right] \qquad (7.3.79)$$

The superscript n refers to the time level and Δt is the time step. It can be seen that the solution for ϕ at time level $n + 1$ can be obtained directly, without the need for iteration, since the matrices are all evaluated at time level n where the conditions are known. It can also be seen that the fluxes J are not averaged over the three time levels but are evaluated directly at time level n. This is to avoid the necessity for iterating on the non-linear vector J^{n+1}. Therefore, assuming two starting values of ϕ, at time levels n and $n-1$, the solution procedure can be started to solve for ϕ^{n+1}. The variables are then updated and repeated use of this relationship enables the distribution of heat and mass transfer potentials to be determined.

7.3.5 Applications

7.3.5.1 Timber drying

To demonstrate the application of complex heat and moisture transfer in practice, a problem of interest to the timber industry is examined [25]. Green timber must be dried before it can be used as a building material and consequently heat and moisture distributions in timber are of considerable interest. The drying of a timber block, 200 mm × 50 mm in section, was chosen as a representative problem and drying was assumed to take place from all surfaces. The problem is shown diagrammatically in Figure 7.3.1, and since the problem is symmetrical only a quarter section need be analysed. A finite element mesh of the resulting 100 mm × 25 mm quarter section is shown in Figure 7.3.2. Drying takes place from surfaces AD and DC, the corner D being subjected to the most fierce drying while corner B, representing the centre of the overall section, will be the slowest to dry. Initial conditions in the body are assumed to be a temperature of 10 °C and a moisture content of 30% (100° moisture potential). The final conditions required are a temperature of 60 °C and a moisture content of 12% (40° moisture potential). Fixed boundary conditions have been applied to this problem.

In the first place a linear analysis of the problem will be considered [25], and values of the thermophysical parameters used are given in Table 7.3.1.

The results are shown in Figures 7.3.3–7.3.5 where the variation in temperature and moisture potential with time are plotted for nodal points 1, 40 and 105, respectively. The units of temperature and moisture potential are both degrees and therefore they can be conveniently shown together. The scale on the time axis is 10 days which is representative of a typical drying schedule.

It can be seen from these results that the variation in temperature is rapid at all nodal points recorded. However, the variation in moisture content is markedly different at the three locations considered. At node 105, near to the drying surfaces of the timber, the variation in moisture potential is also rapid. The dry condition is achieved in a time comparable with the time taken for ambient temperature to be reached. The two processes are therefore obviously coupled in this fast drying region of the wood.

At node 40, which is some distance from the surface, the moisture potential

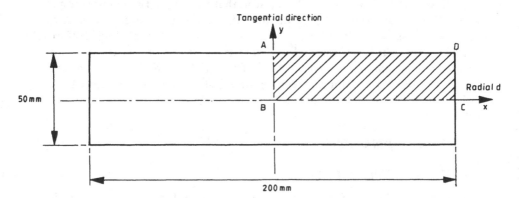

Figure 7.3.1   *Cross-section of timber analysed.*

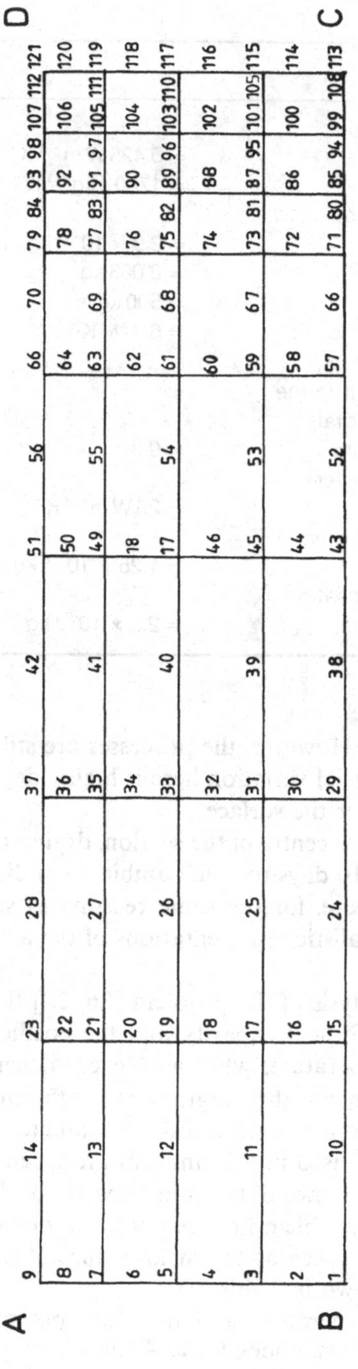

Figure 7.3.2 Finite element discretisation.

Table 7.3.1

Parameter	
Coefficient of thermal conductivity	$= 0.425\,\text{W m}^{-1}\text{K}^{-1}$
Heat capacity	$= 1700\,\text{J kg}^{-1}\text{K}^{-1}$
Coefficient of moisture conductivity	$= 2.9 \times 10^{-9}\,\text{kg}_{\text{moisture}}\,\text{m}^{-1}\,\text{s}^{-1}\,^{\circ}\text{M}^{-1}$
Moisture capacity	$= 0.003\,\text{kg}_{\text{moisture}}\,\text{kg}_{\text{dry body}}^{-1}\,^{\circ}\text{M}^{-1}$
Density	$= 500\,\text{kg m}^{-3}$
Thermogradient coefficient	$= 6.7\,^{\circ}\text{M K}^{-1}$
The ratio of the vapour diffusion coefficient to the coefficient of the total diffusion of moisture	$= 0.3$
Convective heat transfer coefficient	$= 25\,\text{W m}^{-2}\text{K}^{-1}$
Convective mass transfer coefficient	$= 1.25 \times 10^{-6}\,\text{kg m}^{-2}\,\text{s}^{-1}\,^{\circ}\text{M}^{-1}$
Latent heat of vaporisation of water	$= 2.3 \times 10^{6}\,\text{J kg}^{-1}$

variation is less intense. However, the processes are still coupled in this region since the moisture potential variation here is highly dependent on the moisture potential variation nearer the surface.

Finally at node 1, at the centre of the section, drying takes place more slowly, taking a time of nearly 10 days to reach ambient conditions. The processes are however still coupled here, for the same reasons as stated previously. These numerical results are realistic representations of the actual drying process that occurs.

For a non-linear analysis of the problem [26–28] the coefficient of thermal conductivity, the specific heat capacity and the coefficient of convective heat transfer vary with temperature, whereas the coefficient of mass transfer, the phase conversion factor, the thermogradient coefficient and the coefficient of convective mass transfer vary with moisture content.

The numerical values used in this analysis are given in Table 7.3.2.

To examine the significance of the non-linearity of these parameters a comparison run of a linear problem has also been carried out. The values used in that analysis have been taken as the mean values of the variation used in the non-linear work, as shown in Table 7.3.2.

The variation in temperature and moisture potential with time for two representative nodal points, nodes 1 and 40, has been plotted in Figures 7.3.6–7.3.9. Node 1 is chosen to be representative of the central area, whereas node

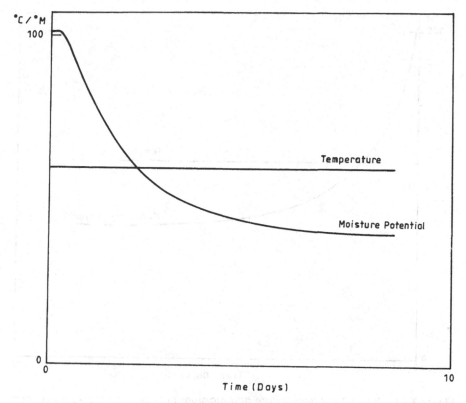

°C / °M

Figure 7.3.3 *Variation of temperature and moisture potential at node 1 for a linear problem.*

40, at the centre of the quarter section, is considered to be representative of the rest of the body. The boundary nodes are all fixed at a specific constant value and there is therefore no variation to be shown for this area.

The four figures (7.3.6–7.3.9) all show two curves, a linear and non-linear result.

An examination of these results shows that for this particular problem only small differences have been observed between the linear and non-linear analyses. This is a surprising result and contrary to generally assumed timber drying behaviour. However, the parameters used were carefully chosen to be representative of timber and therefore these results can be expected to hold true for a typical drying problem. It would, however, be unwise to draw general conclusions on the basis of one series of analyses and further work is needed to substantiate the generality of this finding. In certain cases, for example, the variations in physical parameters may be more severe than those assumed here. Nevertheless it has been shown that both the non-linear and linear analyses can be carried out as demonstrated by the differences observed.

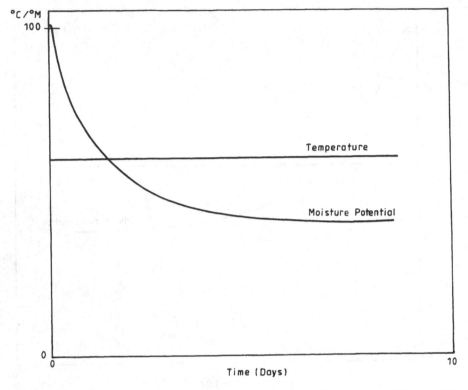

Figure 7.3.4 *Variation of temperature and moisture potential at node 40 for a linear problem.*

7.3.5.2 *Encapsulated electronic circuit example*

The problem demonstrates the effect of a pressure gradient upon the variation in temperature and moisture within a capillary-porous body. Earlier researchers, e.g. Comini and Lewis [23], Thomas [24] and Thomas *et al.* [25], assumed that the pressure within the material under examination was constant and that the pressure gradient, which was assumed to be zero, did not affect the numerical solution. Early studies only examined the effects and interrelationship between temperature and moisture content within the capillary-porous material. The pressure gradient can bring about an additional transfer by means of filtration as well as a diffusional transfer of moisture caused by the presence of a temperature gradient. If it is sufficiently small, then there will be an insignificant effect. For the case of rapid drying, the filtration transfer of moisture is the dominant mode of moisture transfer.

Both the two-degrees-of-freedom and the three-degrees-of-freedom numerical

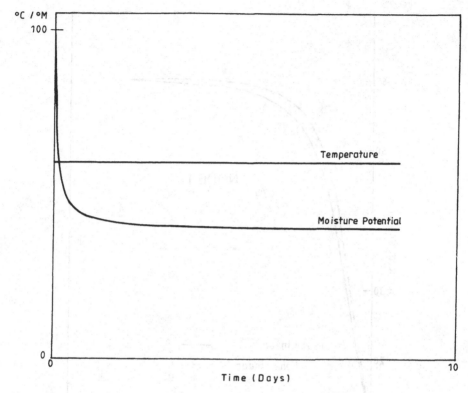

Figure 7.3.5 *Variation of temperature and moisture potential at node 105 for a linear problem.*

Table 7.3.2

Thermophysical parameter	Non-linear problem	Linear problem
k	$0.29\,\mathrm{W\,m^{-1}K^{-1}}$ at $10\,°C$ $0.45\,\mathrm{W\,m^{-1}K}$ at $60\,°C$	$0.35\,\mathrm{W\,m^{-1}K^{-1}}$
c_p	$1163\,\mathrm{J\,kg^{-1}K^{-1}}$ at $10\,°C$ $1405\,\mathrm{J\,kg^{-1}K^{-1}}$ at $60\,°C$	$1284\,\mathrm{J\,kg^{-1}K^{-1}}$
a_m	$0.60 \times 10^{-9}\,\mathrm{m^2\,s^{-1}}$ at 12% m/c $1.54 \times 1^{-9}\,\mathrm{m^2\,s^{-1}}$ at 30% m/c	$1.00 \times 10^{-9}\,\mathrm{m^2\,s^{-1}}$
$\delta*$	$0.01\,°C^{-1}$ at 12% m/c $0.02\,°C^{-1}$ at 30% m/c	$0.02\,°C^{-1}$
ε'	1.0 at 12% m/c 0.1 at 30% m/c	0.3
λ	$2.3 \times 10^6\,\mathrm{J\,kg^{-1}}$	$2.3 \times 10^6\,\mathrm{J\,kg^{-1}}$
ρ	$500\,\mathrm{kg\,m^{-3}}$	$500\,\mathrm{kg\,m^{-3}}$

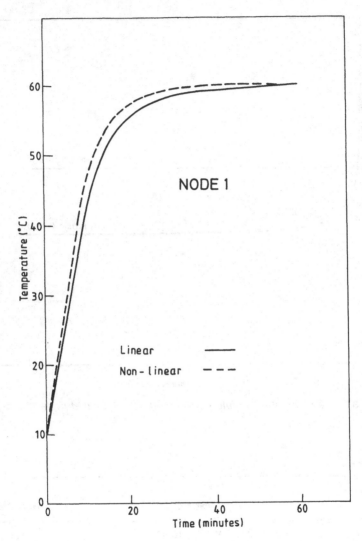

Figure 7.3.6 *Variation of temperature at node 1—linear and non-linear results.*

models were employed to solve this example to demonstrate the effect that the pressure gradient has upon the solution. The three-degree-of-freedom model, which permits the pressure within the body to vary, evaluates the distribution of temperature, moisture potential and pressure throughout the capillary-porous body [42, 43].

The problem to be investigated consists of a container which is designed to prevent the temperature and moisture potential exceeding levels that would age the encapsulated elecronic circuit. For the numerical solution of this problem,

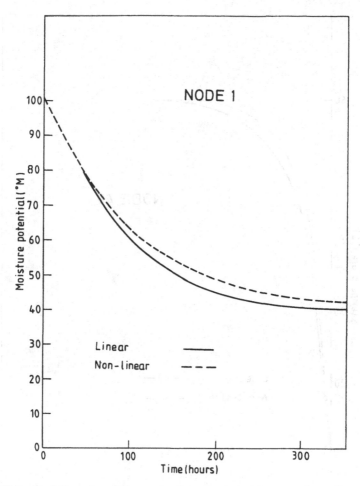

Figure 7.3.7 *Variation of moisture potential at node 1—linear and non-linear results.*

the electronic circuit is not studied; only the distributions of temperature, moisture potential and pressure within the enclosing container are investigated. The cross-section of the enclosing container under examination is shown in Figure 7.3.10. The electronic circuit, which is not shown in Figure 7.3.10, lies along the x-axis and symmetrical about the y-axis within the silicon gel material.

The container consists of an epoxy resin casing with an injected silicon gel core which surrounds and protects the electronic circuit. The epoxy resin casing and silicon gel interior are separated by a moisture barrier, which lies parallel to the x-axis, and is designed to prevent excessive moisture migration in the y-direction from the epoxy casing to the silicon gel centre. This is achieved numerically by using a comparatively small value for the coefficient of moisture

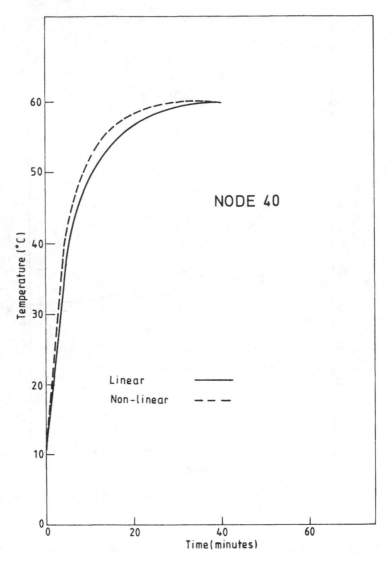

Figure 7.3.8   *Variation of temperature at node 40—linear and non-linear results.*

conductivity, k_m. Although the moisture barrier is designed to prevent an excessive moisture migration, it is not intended to be an insulator, and therefore heat is conducted freely across the barrier film between the epoxy resin and silicon gel. The material properties, which are assumed to be constant, required for the numerical model were obtained from [42] and are shown in Table 7.3.3.

The problem under investigation is symmetrical about the y-axis; hence only

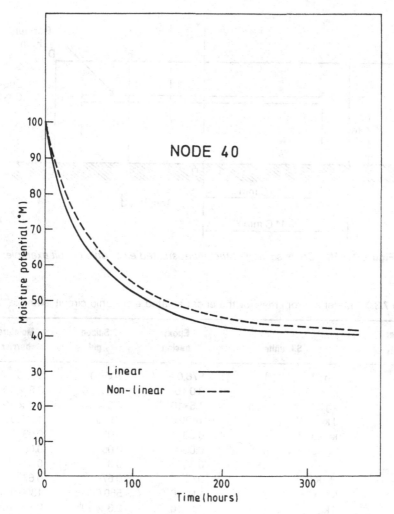

Figure 7.3.9   *Variation of moisture potential at node 40—linear and non-linear results.*

the region ABCD is analysed in order to reduce the solution time. The initial conditions throughout the domain of interest are temperature 30 °C, moisture potential 13.5 °M and pressure 0 kN m⁻². Along faces AD and CD a flux-type boundary condition, with a ramp loading, was applied to the temperature and the moisture potential, whilst for the pressure term a fixed point boundary condition was applied, which was also with a ramp loading. The boundary conditions increased linearly from the initial conditions to the encapsulated electronic circuit's steady-state condition, temperature 80 °C, moisture potential 28.5 °M and pressure 100 kN m⁻², over a period of 2 h. The ramp loadings of

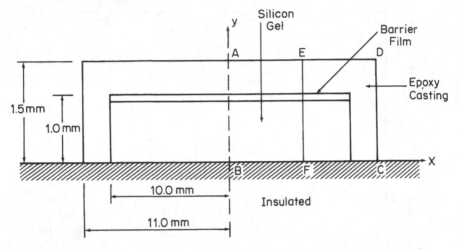

Figure 7.3.10 *Cross-section of the encapsulated electronic circuit example.*

Table 7.3.3 Material properties for the encapsulated electronic circuit example.

Material property	S.I. units	Epoxy casing	Silicon gel	Moisture barrier
k_q	$J\,h^{-1}\,m^{-1}\,K^{-1}$	576.0	7055.0	108.0
k_m	$m\,h^{-2}$	$3.0\,10^{-6}$	4.0×10^{-7}	2.8×10^{-9}
k_p	$kg\,h^{-1}\,m^{-1}\,N^{-1}\,m^{-1}$	$1.5*10^{-8}$	2.0×10^{-7}	1.4×10^{-9}
C_q	$J\,kg^{-1}\,K^{-1}$	1400.0	246.0	1210.0
C_m	$kg\,kg^{-1}\,{}^{\circ}M^{-1}$	0.03	0.03	0.03
C_p	$kg\,kg^{-1}\,N^{-1}\,m^{-2}$	0.05	0.05	0.05
ε		0.3	0.3	0.3
δ	${}^{\circ}M\,K^{-1}$	0.67	0.67	0.67
ρ_0	$kg\,m^{-3}$	1170.0	550.0	1300.0
λ	$J\,kg^{-1}$	$2.3*10^{-6}$	2.3×10^6	2.3×10^6

the fixed point pressure boundary condition and the ambient temperature and moisture potential are shown in Figure 7.3.11. Along sides AB and BC it was assumed that there was a non-conducting boundary condition and that these faces were insulated.

At point D on Figure 7.3.11 the migration of temperature, moisture potential and pressure is two-dimensional owing to the nature of the boundary conditions along faces AD and CD. However, as we approach face AB the migration of temperature, moisture potential and pressure becomes one-dimensional in the y-direction. Close to the y-axis there is no flow in the x-direction because the

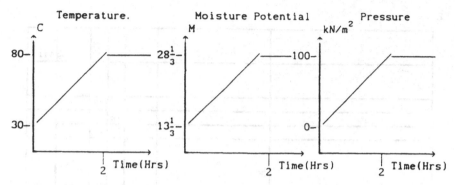

Figure 7.3.11   *Ramp loading boundary conditions.*

effect of the applied boundary condition along face CD diminishes as we approach the *y*-axis. Therefore, if we are sufficiently far from face CD it can be assumed that the flow is one-dimensional and that there is a non-conducting boundary, in the *x*-direction, along an imaginary internal face EF. The area of interest, and that for which the temperature, moisture potential and pressure distributions are calculated, is region EFCD, where two-dimensional flow occurs. In region ABFE the temperature, moisture potential and pressure values are equal to those along face EF for a given *y*-value. No change occurs within this region along the *x*-axis. One-dimensional flow would have occurred naturally within this region during the numerical solution process, but to reduce the analysis time, only the region where two-dimensional flow occurs was solved. The area that was discretised was area EFCD and the finite element mesh used for the numerical solution is shown in Figure 7.3.12. Close to point D, the elements are small owing to the fact that this is the area where the gradients of temperature, moisture potential and pressure are expected to be highest due to the proximity of the faces along which the boundary conditions are applied.

Figures 7.3.13–7.3.17 show the variation of temperature with time, Figures 7.3.18–7.3.22 show the variation of moisture potential with time and Figures 7.3.23–7.3.27 show the variation of pressure with time for nodes 15, 36, 78, 99 and 141. On each figure the results are plotted from the two-degrees-of-freedom system, where only the temperature and moisture potential distributions within the capillary–porous body are calculated, and the three-degrees-of-freedom system, where the temperature, moisture potential and pressure distributions are calculated, and a comparison is made between the solutions from both numerical models. The three-degrees-of-freedom model permits the pressure term to vary within the capillary–porous body, whilst the two-degrees-of-freedom model assumes that the pressure term is constant throughout the whole of the domain of interest and that the pressure gradient is zero everywhere within the body. Therefore, the pressure versus time graphs only show the variation in pressure with time for the three-degrees-of-freedom numerical model.

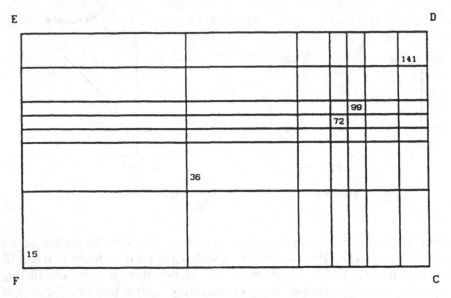

Figure 7.3.12 *Finite element discretisation of the encapsulated electronic circuit example.*

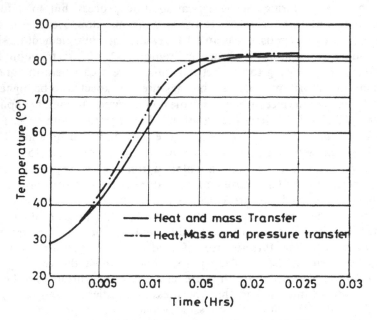

Figure 7.3.13 *Temperature vs time at node 15.*

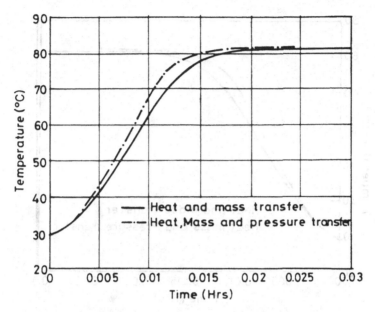

Figure 7.3.14   *Temperature vs time at node 36.*

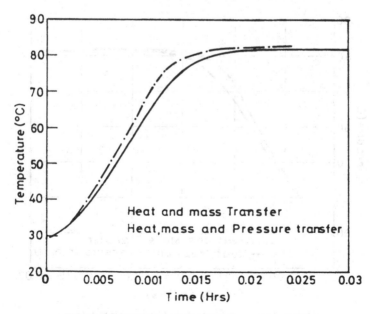

Figure 7.3.15   *Temperature vs time at node 78.*

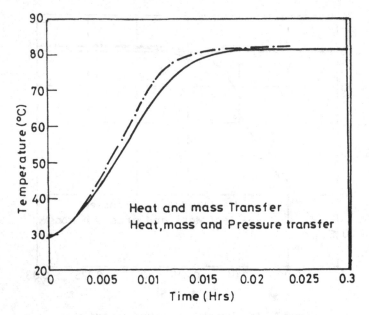

Figure 7.3.16 *Temperature vs time at node 99.*

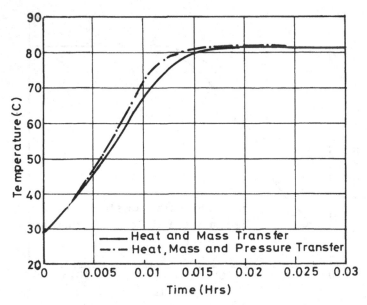

Figure 7.3.17 *Temperature vs time at node 141.*

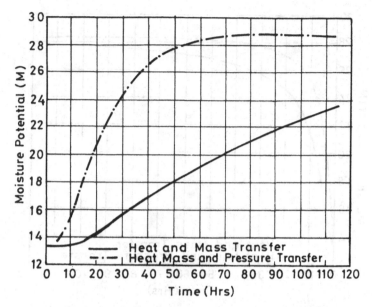

Figure 7.3.18   *Moisture potential vs time at node 15.*

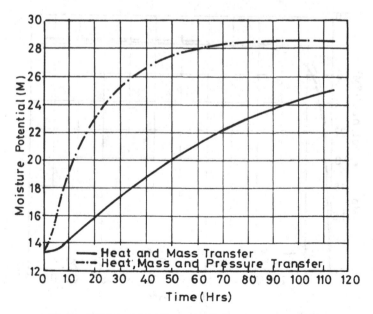

Figure 7.3.19   *Moisture potential vs time at node 36.*

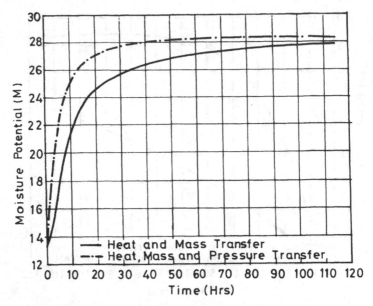

Figure 7.3.20 *Moisture potential vs time at node 78.*

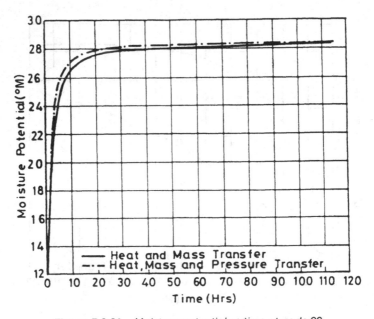

Figure 7.3.21 *Moisture potential vs time at node 99.*

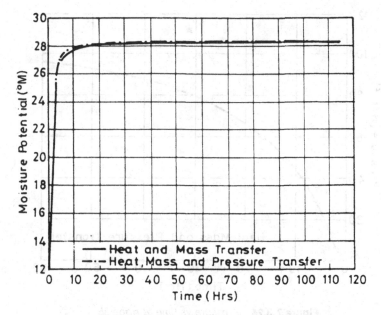

Figure 7.3.22 *Moisture potential vs time at node 141.*

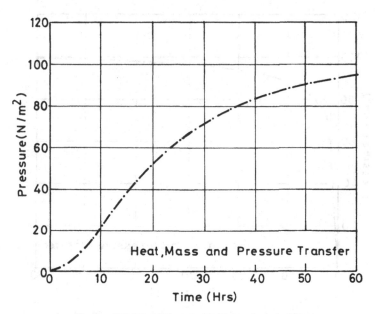

Figure 7.3.23 *Pressure vs time at node 15.*

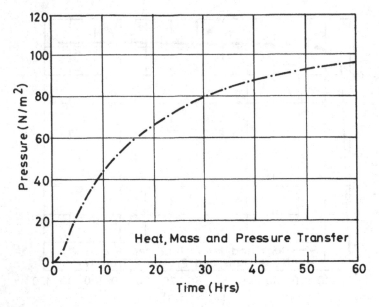

Figure 7.3.24 *Pressure vs time at node 36.*

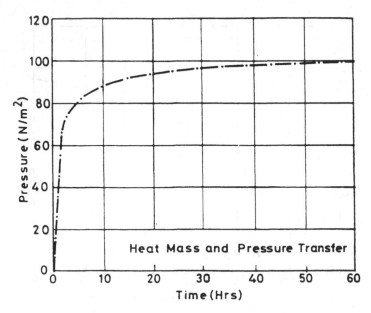

Figure 7.3.25 *Pressure vs time at node 78.*

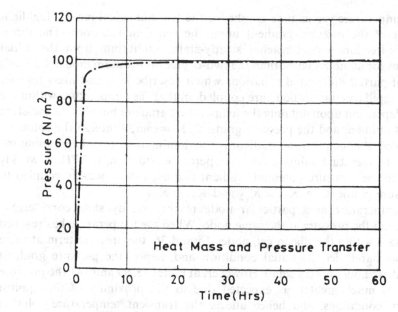

Figure 7.3.26   *Pressure vs time at node 99.*

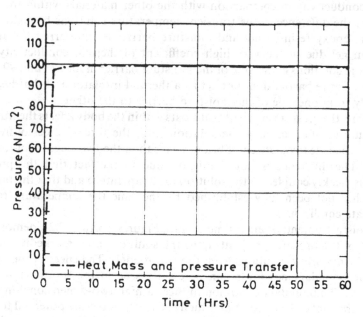

Figure 7.3.27   *Pressure vs time at node 141.*

The temperature versus time graphs for the two numerical models highlight the effect of the pressure gradient upon the transient solutions. The three-degrees-of-freedom model reaches steady-state equilibrium from the initial conditions for all nodes faster than the two-degrees-of-freedom model. Luikov's system of partial differential equations which describe heat and mass transfer within a capillary–porous body are coupled, in that the temperature within the body is dependent upon not only the temperature gradient but also the moisture potential gradient and the pressure gradient. However, although the solutions are coupled, the temperature gradient is the dominant term is determining the transient temperature solution. The temperature equation, (7.3.71), is weakly coupled to the moisture potential gradient and even more weakly coupled to the pressure gradient, i.e. $K_{11} \gg K_{12}$ and $K_{11} \gg K_{13}$.

The temperature at a particular node reaches steady state considerably quicker than the pressure at the same node. When the temperature has reached steady state within the silicon gel, nodes 15 and 36, the pressure term at these nodes has barely left its initial condition and, hence, the pressure gradient around these nodes is very small. However, at nodes 78, 99 and 141 the pressure gradient is much greater, as expected due to the proximity of the applied boundary conditions, and hence affects the transient temperature solution thereby causing the difference in solutions obtained from the two numerical models. The difference in the temperature solutions at nodes 15 and 36 cannot be attributed to the effect of the pressure gradient, which is very small in this region, but due to the fact that the silicon gel has a very high coefficient of thermal conductivity in comparison with the other materials within the body. Therefore, the difference in the transient temperature solutions brought about within the epoxy resin casing and moisture barrier is transferred through to the silicon gel due to the very high coefficient of thermal conductivity. The temperature solutions either side of the moisture barrier, nodes 78 and 99, show that the moisture barrier does not act as a thermal insulator and that heat can pass freely from one side of the moisture barrier to the other.

Therefore, the pressure gradient that exists within the body affects the transient temperature solution causing the solution from the three-degrees-of-freedom model to reach steady-state equilibrium faster than the two-degrees-of-freedom solution. The difference is not significant due to the fact that the pressure gradient is weakly coupled to the solution for temperature and that the pressure gradient has not been fully established by the time the temperature reaches steady-state equilibrium.

The moisture potential versus time graphs, Figures 7.3.18–7.3.22, demonstrate the effect of the pressure gradient upon the solution more markedly than the transient temperature solutions from the two models. The effect of the applied convective boundary condition can be clearly seen in the solutions at nodes 141 and 99. The boundary condition induces a high two-dimensional moisture potential gradient close to point D, which causes the moisture potential to reach steady state quickly at these nodes. The pressure gradient does not significantly affect the solutions at these nodes because the moisture potential gradient is the dominant term in the moisture potential equation (7.3.72), i.e. $K_{22} \gg K_{23}$. However, as we proceed away from point D to the interior of the problem, nodes 78, 36 and 15, the moisture potential gradient decreases and the pressure gradient plays a more prominent role. The effect of this can be seen in the very marked difference in the transient moisture potential solutions at nodes 78, 36 and 15 obtained from the two numerical methods. The presence of the pressure gradient effects a transfer of moisture by filtration in addition to the transfer of moisture by diffusion. The two-degrees-of-freedom model assumes that the pressure gradient is zero throughout the whole of the domain of interest, therefore, no filtration transfer of moisture occurs. Moisture transfer takes place by means of diffusion only. It is the presence of the pressure gradient, and hence the filtration transfer of moisture, that causes the transient moisture potential solution for the three-degrees-of-freedom model to reach steady-state equilibrium very much more quickly than the moisture potential solution obtained from the two-degrees-of-freedom model. The effect of the presence of a moisture barrier can be seen in the moisture potential solutions at nodes 99 and 78. The moisture potential at node 99 reaches steady state quicker than node 78, because the moisture barrier inhibits the transfer of moisture, by diffusion or filtration, across it and acts as a moisture insulator.

This numerical example demonstrates the significant effect of the pressure gradient upon the solutions for temperature and moisture potential. The marked difference is caused by an additional transfer of moisture by means of filtration and highlights the fact that the pressure gradient within a capillary–porous body cannot be assumed to be zero and must be taken into account for a problem involving rapid drying.

7.3.5.3 Application in frost heave in freezing soil and freeze drying

The coupled heat and mass transfer partial differential equations as derived by Luikov are also used to analyse freezing problems in porous media. A numerical model is developed to depict the frost heave mechanism in a freezing soil system, the results of which compare well with those achieved from field experiments [44–46]. A numerical modelling of freeze drying of a 3.6% coffee solution showed a good correlation with experimental results [47].

7.4 SHRINKAGE STRESS DEVELOPMENT

The basis of the method of calculating shrinkage stress development is similar to the method employed in Section 7.2 for thermal stress development. Details of the overall procedure need not therefore be repeated here. Shrinkage stress analysis is illustrated in this section for the problem of timber drying considered previously.

Timber drying schedules are designed to minimise energy costs by keeping drying time low. This requires the fastest rate of drying possible. However, if too fast a rate of drying is employed, damage occurs to the timber being dried and the gains achieved from lower energy bills are offset by the costs incurred in a lower quality product. There is therefore a need to find an optimum drying rate that minimises operational costs without incurring wastage costs. To this end, therefore, a numerical solution of shrinkage stresses arising out of heat and mass transfer distributions is a useful predictive tool.

As stated previously, three constitutive relationships will be examined in this section; linear elastic, elasto-plastic and viscoelastic. These models have been used to date to examine a number of problems of engineering significance [10, 11, 34–38]. For the analysis of timber problems, however, an orthotropic model is required and this section concentrates on the application of the models to an orthotropic material. The basic details of the methods used are, however, first explained before proceeding with the finite element solution procedures and applications.

7.4.1 Elastic stress/strain relationships

For an orthotropic material, elastic equations can be written as

$$\varepsilon_x = \frac{\sigma_x}{E_x} - v_{yx}\frac{\sigma_y}{E_y} - v_{zx}\frac{\sigma_z}{E_z}$$

$$\varepsilon_y = - v_{xy}\frac{\sigma_x}{E_x} + \frac{\sigma_y}{E_y} - v_{zy}\frac{\sigma_z}{E_z}$$

$$\varepsilon_z = - v_{xz}\frac{\sigma_x}{E_x} - v_{yz}\frac{\sigma_y}{E_y} + \frac{\sigma_z}{E_z}$$ (7.4.1)

$$\gamma_{xy} = \frac{1}{G_{xy}}\tau_{xy}$$

where E_x, E_y and E_z are Young's modulii in the axes directions, v_{xy}, etc. are the respective Poisson's ratios and G_{xy} is now an independent variable.

Figure 7.4.1 shows the idealisation of wood as a cylindrically orthotropic material and shows the relationship between the cartesian co-ordinate axes and the principal directions of wood usually referred to in the literature, i.e. radial, tangential and longitudinal.

In order to reduce the number of variables in the equations, advantage can be taken of the relationships

$$v_{xy}/E_x = v_{yx}/E_y$$
$$v_{xz}/E_x = v_{zx}/E_z$$ (7.4.2)
$$v_{yz}/E_y = v_{zy}/E_z$$

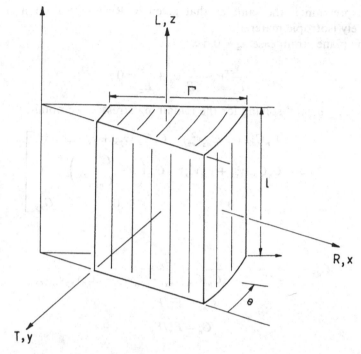

Figure 7.4.1 *Idealisation of wood as a cylindrically orthotropic material.*

Therefore it can be seen that if the above equations are expressed in matrix form as

$$\varepsilon^e = D^{-1}\sigma \tag{7.4.3}$$

D is a symmetric matrix.

During the drying process the strength properties of timber are greatly increased and therefore it must be remembered that the values of these elasticity parameters change throughout the analysis as functions of moisture content.

For plane stress situations, $\sigma_z = 0$, and solving for the above stresses, we obtain the matrix D as

$$D = \begin{bmatrix} C_n C_d & C_n C_d v_{yx} & 0 \\ C_n C_d v_{yx} & C_d & 0 \\ 0 & 0 & G_{xy} \end{bmatrix} \tag{7.4.4}$$

where

$$C_n = E_x/E_y \tag{7.4.5}$$

and

$$C_d = \frac{E_y}{1 - C_n v_{yx}^2} \tag{7.4.6}$$

This expression is the same as that given in Reference 1 for a stratified, transversely isotropic material.

For the plane strain case, $\varepsilon_z = 0$, i.e.

$$-\frac{v_{zx}}{E_z}\sigma_x - \frac{v_{yz}}{E_y}\sigma_y + \frac{\sigma_z}{E_z} = 0 \tag{7.4.7}$$

Eliminating σ_z and solving for the remaining stresses, we obtain

$$\mathbf{D} = \begin{bmatrix} C_n C_{dz}(1 - C_{nz}v_{yz}^2) & C_n C_{dz}(v_{yx} + v_{yz}v_{zx}) & 0 \\ C_n C_{dz}(v_{yx} + v_{yz}v_{zx}) & C_{dz}\left(1 - \frac{C_n}{C_{nz}}v_{zx}^2\right) & 0 \\ 0 & 0 & G_{xy} \end{bmatrix} \tag{7.4.8}$$

where

$$C_{dz} = \frac{E_y}{1 - C_{xz}v_{yz}^2 - (C_n/C_{nz})v_{zx}^2 - C_n v_{yx}^2 - 2C_n v_{yx}v_{yz}v_{zx}} \tag{7.4.9}$$

$$C_{nz} = E_z/E_y \tag{7.4.10}$$

$$C_n = E_x/E_y \tag{7.4.11}$$

7.4.2 Plastic yield criterion

The yield criterion chosen to represent the plastic behaviour of an orthotropic material is based on the work of Hill [29] who based his theory on the hypothesis that the simplest yield criterion for anisotropic material is one that reduces to Von Mises' law when the anisotropy is vanishingly small. Therefore if the yield criterion is assumed to be quadratic in the stress components, it must be of the form

$$1 = K_1(\sigma_y - \sigma_z)^2 + K_2(\sigma_z - \sigma_x)^2 + K_3(\sigma_x - \sigma_y)^2$$
$$+ 2K_4\tau_{yz}^2 + 2K_5\tau_{zx}^2 + 2K_6\tau_{xy}^2 \tag{7.4.12}$$

where K_1, K_2, K_3, K_4, K_5 and K_6 are parameters characteristic of the state of anisotropy. Linear terms are not included since it is assumed that there is no Bauschinger effect. Quadratic terms in which any one shear stress occurs linearly are rejected in view of symmetry restriction. Finally, only the differences of the normal components can appear if it is assumed that the superimposition of a hydrostatic stress does not influence yielding.

If X, Y and Z are the tensile yield stresses in the principal directions of anisotropy, then for the case of uniaxial tension

$$\sigma_x = X = \sigma_1 \qquad \sigma_y = \sigma_z = \tau_{yz} = \tau_{zx} = \tau_{xy} = 0$$

Substituting into (7.4.12) yields

$$K_2 X^2 + K_3 X^2 = 1$$

i.e.

$$\frac{1}{X^2} = K_2 + K_3 \tag{7.4.13}$$

Similarly, we may write

$$\frac{1}{Y^2} = K_3 + K_1 \tag{7.4.14}$$

$$\frac{1}{Z^2} = K_1 + K_2$$

Rearranging these, we can obtain expressions for K_1, K_2 and K_3 as follows:

$$2K_1 = \frac{1}{Y^2} + \frac{1}{Z^2} - \frac{1}{X^2}$$

$$2K_2 = \frac{1}{Z^2} + \frac{1}{X^2} - \frac{1}{Y^2} \tag{7.4.15}$$

$$2K_3 = \frac{1}{X^2} + \frac{1}{Y^2} - \frac{1}{Z^2}$$

If YZ, ZX and XY are the yield stresses in shear with respect to the principal axes of anisotropy, then

$$2K_4 = \frac{1}{YZ^2} \qquad 2K_5 = \frac{1}{ZX^2} \qquad 2K_6 = \frac{1}{XY^2} \tag{7.4.16}$$

From these expressions we can see that K_4, K_5, and K_6 are positive but one of K_1, K_2 or K_3 can be negative. Equation (7.4.12) can be written alternatively as

$$\begin{aligned}
[(K_2 + K_3)\sigma_x^2 - 2K_3\sigma_x\sigma_y + (K_1 + K_3)\sigma_y^2 + 2K_6\tau_{xy}^2] \\
- 2(K_2\sigma_x + K_1\sigma_y)\sigma_z + 2(K_4\tau_{yz}^2 + K_5\tau_{zx}^2) + (K_1 + K_2)\sigma_z^2 = 1
\end{aligned} \tag{7.4.17}$$

To describe fully therefore the plastic state of anisotropy in a material, we need to know the values of the six independent yield stresses. They must be considered as functions of changing moisture content and temperature. For two-dimensional problems, knowledge of the four yield stresses, X, Y, Z and XY, are sufficient to define the problem.

7.4.3 Viscoelastic behaviour

The viscoelastic behaviour of timber during drying is analysed by means of an approach proposed by Srinatha [30] whereby temperature and moisture content influence on viscoelasticity is described by a time–temperature and moisture content equivalence hypothesis. In a thermoviscoelastic problem, for example, the appearance of temperature as a variable is removed by shifting the real time based on a time–temperature equivalence hypothesis [31]. This approach is extended for a drying problem to include both time–temperature and time–moisture content equivalence.

Assuming the validity of the above hypothesis, the real time, t, is related to temperature, T, and moisture content, M, so that single-shifted time can be used in place of the three variables t, T and M. For isohygrothermal situations, therefore, the following relations exist, for example:

$$E(T, m, t) = E(T_0, m_0, \xi) \qquad \xi = \Phi(T, m)t \qquad (7.4.18)$$

where T_0 and m_0 are reference temperature and moisture content, respectively, and $\Phi(T, m)$ is the shifting function.

The superimposition principle used above states that a constant change in temperature and moisture content in the body affects its mechanical responses within only a uniform distortion of the time scale. The behaviour of materials at high temperature and moisture content is similar to those at low temperature and low moisture content. Materials which could be characterized by this superimposition were defined as 'thermohygrorheologically simple' materials [30].

For non-isohygrothermal situations, the shifted time is given by

$$\xi = \int_0^t \Phi\{T(x, t), m(x, t)\} \, dt$$

The shifting function, $\Phi(T, m)$, is a monotonically increasing function that satisfies $\Phi(T_0, m_0) = 1$, $\Phi(T, m) > 0$, $\partial \Phi / \partial t > 0$, $\partial \Phi / \partial m > 0$. The shift function remains the same for responses in shear and tension and this follows from the conditions of compatibility imposed in defining a thermohygrorheologically simple solid.

To determine the shifting function, the following experimental procedure may be adopted:

(1) Conduct creep or relaxation tests on the material specimen at an isothermal temperature at various moisture content levels. In the case of timber, these levels should correspond to the actual moisture content levels during drying.
(2) For a constant moisture content, conduct a creep test at various temperature levels.
(3) Using the creep curves from (1) and (2), a master creep or relaxation curve can be constructed. This can then be used to determine the shift function $\Phi(T, m)$.

In general terms, the constitutive relationship for anisotropic viscoelasticity has the form

$$\sigma_{ij}(t) = c_{ij}^{kl}(0)\varepsilon_{kl}(t) - \int_0^t \frac{\partial}{\partial t'}(c_{ij}^{kl}(t-t'))\varepsilon_{kl}(t)\,dt' \tag{7.4.19}$$

where the initial response has been made explicit. c_{ij}^{kl} is the tensor of material responses, which has 21 independent functions. For the general case of orthotropy, this reduces to five independent time functions. To ensure compatibility with an elasticity formulation the constitutive equation was expressed in terms of four compliance functions and the in-plane rigidity modulus.

To specify the precise form of the constitutive relationship used in this work, the following assumptions have been made:

(1) The material is linear and homogeneous.
(2) The hypothesis of a thermohygrorheologically simple solid is valid. The real time, t, can be transformed into shifted time through the shift function, which is determined in terms of the temperature and moisture content.
(3) The material compliance functions are expressed in terms of a Prony–Dirichlet series in time, e.g.

$$c_{11} = c_{110} + \sum_{i=1}^{n} c_{11i}\exp(-t/\zeta_i) \tag{7.4.20}$$

where ζ_i are the relaxation times.
(4) The heat and mass transfer taking place during the drying of timber is independent of the viscoelastic stress response.

With the last assumption, the heat and mass transfer problem may be solved concurrently with the stress analysis in the same way as the elastic and elastoplastic analyses are performed.

The constitutive equations may then be expressed as

$$\sigma = \mathbf{D}(0)\varepsilon - \int_0^t \frac{\partial}{\partial\xi'}[\mathbf{D}(\xi-\psi')]\varepsilon(t)\,d\xi'$$

$$- \mathbf{D}(0)(\varepsilon_T + \varepsilon_m) + \int_0^t \frac{\partial}{\partial\xi'}(\mathbf{D}(\xi-\xi'))(\varepsilon_T + \varepsilon_m)\,d\xi' \tag{7.4.21}$$

where ξ is the shifted time, given by

$$\xi = \int_0^t \Phi(T,m)\,dt \tag{7.4.22}$$

and ξ' is the shifted time prior to ξ. In equation (7.4.21), the initial elastic response has been made explicit:

$$D(0) = \begin{bmatrix} C_{11}^{(0)} & C_{12}^{(0)} & 0 \\ C_{21}^{(0)} & C_{22}^{(0)} & 0 \\ 0 & 0 & G_{12}^{(0)} \end{bmatrix} = \text{initial elastic matrix} \tag{7.4.23}$$

$$D(\xi - \xi)' = \begin{bmatrix} C_{11}(\xi - \xi)' & C_{12}(\xi - \xi)' & 0 \\ C_{21}(\xi - \xi)' & C_{22}(\xi - \xi)' & 0 \\ 0 & 0 & G_{12}(\xi - \xi)' \end{bmatrix} = \text{viscoelastic matrix}$$

(7.4.24)

ε_T and ε_m are the initial strain vectors due to temperature and moisture content changes, respectively. Unlike an isotropic problem, it is not possible to extract the time-varying part from the constant part in the orthotropic problem [30].

7.4.4 Finite element formulation

The finite element solution method for the elastic and elastoplastic cases has been discussed in Section 7.2 and will not be repeated here. The finite element solution of viscoelastic behaviour is however new to this section and will be discussed.

If the constitutive equation (7.4.21) is introduced into the statement of virtual work, the following finite element equations are obtained:

$$\mathbf{K}'\delta^e - \int_{\Omega^e} \int_0^t \mathbf{B}^T \frac{\partial}{\partial \xi'} [\mathbf{D}(\xi - \xi')] \varepsilon \, d\xi' \, d\Omega$$

$$+ \int_{\Omega^e} \int_0^t \mathbf{B}^T \frac{\partial}{\partial \xi'} [\mathbf{D}(\xi - \xi')] (\varepsilon_T + \varepsilon_m) \, d\xi' \, d\Omega = \mathbf{F}^e + \mathbf{F}^e_{Tm} \quad (7.4.25)$$

where

$$\mathbf{K}' = \int_\Omega \mathbf{B}^T [\mathbf{D}(0)] \mathbf{B} \, d\Omega = \text{initial elastic stiffness matrix} \quad (7.4.26)$$

and

$$\mathbf{F}^e_{Tm} = \int_\Omega \mathbf{B}^T [\mathbf{D}(0)] (\varepsilon_T + \varepsilon_m) \, d\Omega = \text{load vector due to changes in temperature}$$
$$\text{and moisture content, respectively} \quad (7.4.27)$$

The above equations contain two hereditary integrals. These are Volterra integrals of the first kind in time. To simplify their evaluation they are approximated by algebraic expressions according to the method proposed by Hopkins and Hamming [32] and Lee and Rogers [33]. To illustrate the approximations involved, the first hereditary integral will be considered, omitting the volume integral for simplicity. The matrix $\mathbf{B}^T$ is also omitted since it is a function of only the space co-ordinates. The resulting expression is

$$\oint \frac{\partial}{\partial \xi'} [\mathbf{D}(\xi - \xi')] \varepsilon \, d\xi' \, d\Omega \quad (7.4.28)$$

If the time scale is divided into intervals by time values, t_i, $i = 1, 2, \ldots, k$ such that $t_1 = 0$ and $t_k = t$, equation (7.4.28) can be written in the form

$$\oint_0^t \frac{\partial}{\partial \xi'} [\mathbf{D}(\xi - \xi')] \varepsilon \, d\xi' = \int_{t_i}^{t_k} \frac{\partial}{\partial \xi'} [\mathbf{D}(\xi - \xi')] \varepsilon \, d\xi'$$

$$= \sum_{i=1}^{k-1} \int_{t_i}^{t_{i+1}} \frac{\partial}{\partial \xi'} [\mathbf{D}(\xi_k - \xi')] \varepsilon \, d\xi' \qquad (7.4.29)$$

Each of the integrals in equation (7.4.29) can be approximated by the following trapezoidal finite difference scheme:

$$\int_{t_i}^{t_{i+1}} \frac{\partial}{\partial \xi'} [\mathbf{D}(\xi_k - \xi')] \varepsilon \, d\xi' \simeq \tfrac{1}{2} [\mathbf{D}(\xi_k - \xi_{i+1}) - \mathbf{D}(\xi_k - \xi_i)] \varepsilon^*(t_i) \quad (7.4.30)$$

where

$$\varepsilon^*(t_i) = \tfrac{1}{2}(\varepsilon(t_i) + \varepsilon(t_{i+1})) \qquad (7.4.31)$$

Substituting equation (7.4.30) into (7.4.31) and expanding for $t_i = k - 1$, the following approximation is obtained:

$$\int_0^t \frac{\partial}{\partial \xi'} [\mathbf{D}(\xi - \xi')] \varepsilon \, d\xi' = \sum_{i=1}^{k-2} \tfrac{1}{2} [\mathbf{D}(\xi_k - \xi_{i-1}) - \mathbf{D}(\xi_k - \xi_i)] \varepsilon^*(t_i)$$

$$+ \tfrac{1}{2} [\mathbf{D}(0) - \mathbf{D}(\xi_k - \xi_{k-1})] \varepsilon(t_{k-1})$$

$$+ \tfrac{1}{2} [\mathbf{D}(0) - \mathbf{D}(\xi_k - \xi_{k-1})] \varepsilon(t_k) \qquad (7.4.32)$$

The finite difference scheme is extremely stable and the error propagation is very small [32, 33]. The time marching scheme reduces to such a simple form because the integration limits in the Volterra integral equations are from 0 to t and the solution of the equilibrium equations at $t = 0$ corresponds with those of instantaneous elasticity.

At time t_k all the displacement vectors up to t_{k-1} are known. Thus, using equation (7.4.32), the first integral in the equilibrium equations (7.4.21) may be written in the form

$$\int_{\Omega^e} \int_0^t \mathbf{B}^T \frac{\partial}{\partial \xi'} [\mathbf{D}(\xi - \xi')] \varepsilon \, d\xi' \, d\Omega = -\mathbf{F}_{m1}^e(t_k) + \mathbf{K}_2 \delta^e(t_k) \qquad (7.4.33)$$

where

$$\mathbf{K}_2 = \int_\Omega \mathbf{B}^T [\tfrac{1}{2} \{\mathbf{D}(0) - \mathbf{D}(\xi_k - \xi_{k-1})\}] \mathbf{B} \, d\Omega \qquad (7.4.34)$$

and

$$\mathbf{F}_{m1}^e(t_k) = -\int_{\Omega^e} \int_0^t \mathbf{B}^T \sum_{i=1}^{k-2} \tfrac{1}{2} \{\mathbf{D}(\xi_k - \xi_{i+1}) - \mathbf{D}(\xi_k - \xi_i)\} \varepsilon^*(t_i)$$

$$- \int_{\Omega^e} \int_0^t \mathbf{B}^{T} \tfrac{1}{2} \{\mathbf{D}(0) - \mathbf{D}(\xi_k - \xi_{k-1})\} \varepsilon(t_{k-1}) \qquad (7.4.35)$$

To evaluate the summation in equation (7.4.35), it is necessary to store the strains at all the time steps up to time t_{k-1}. This expensive computer storage and retrieval can be avoided, however, by the development of a suitable recurrence relationship. Most viscoelastic materials have discrete relaxation spectrums and exponentially decreasing relaxation times. It is for this reason that the material compliance functions have been expressed in terms of a Prony–Dirichlet series.

Hence substituting from (7.4.20) into the summation in equation (7.4.35) and interchanging the summations, the following recurrence relationship is obtained:

$$\sum_{i=1}^{k-1} \tfrac{1}{2}\{\mathbf{D}(\xi_k - \xi_{i+1}) - \mathbf{D}(\xi_k - \xi_i)\}\varepsilon^*(t_i) = \sum_{j=1}^{m} \tfrac{1}{2}\mathbf{D}_j\mathbf{q}_{j,k} \qquad (7.4.36)$$

where

$$\mathbf{q}_{j,k} = \exp(-(\xi_k - \xi_{k-1})/\zeta_i)[\{(1 - \exp(-\xi_{k-1} + \xi_{k-2})/\zeta_j)\varepsilon^*(t_{k-2})\} + \mathbf{q}_{j,k-1}] \qquad (7.4.37)$$

For the first two time steps, $q_{j,1}$ and $q_{j,2}$ are zero vectors. Thus, to evaluate $q_{j,k}$, two immediately past strain vectors and the recurrence vector $q_{j,k-1}$ are required. Similar recurrence relations may be derived for the other hereditary integrals in equation (7.4.25). On substituting equation (7.4.36) into (7.4.32), the memory loads are obtained. Finally, the set of equilibrium equations at time t'_k, which are now algebraic equations, may be expressed symbolically as

$$\{\mathbf{K}1 - \mathbf{K}2\}\delta^e(t_k) = \mathbf{F}^e + \mathbf{F}^e_{\text{Tm}} + \mathbf{F}^e_{\text{ML}} + \mathbf{F}^e_{\text{MTm}} = \mathbf{F}^e_{\text{T}} \qquad (7.4.38)$$

where $\mathbf{F}^e_{\text{ML}}$ = memory load due to strains and $\mathbf{F}^e_{\text{MTm}}$ = memory loads due to changes in temperature and moisture content.

The assembly and solution of equations (7.4.38) subject to prescribed boundary displacements follows well-established procedures. After solution, the stresses are obtained by substituting the strains into the constitutive equation (7.4.21).

If required, an alternative form of the equilibrium equations may be derived in conjunction with an iterative scheme, so that the stiffness matrices are maintained constant for all time steps. In such a scheme, a forward elimination of matrices is carried out only once, while for the remaining time steps only a backward substitution will be required.

The alternative equilibrium equations are

$$\mathbf{K}1\delta^e_{i+1}(t_k) = \mathbf{F}^e_{\text{T}} + \mathbf{K}2\delta^e_i(t_k) \qquad (7.4.39)$$

where i is the iteration number and

$$\delta^e_{i+1}(t_k) = \delta^e(t_{k-1})$$

The iteration is terminated when, for non-zero terms in $\delta^e_{i+1}(t_k)$,

$$\left\| \frac{\delta^e_{i+1}(t_k) - \delta^e_i(t_k)}{\delta^e_{i+1}(t_k)} \right\| < \text{eps} \qquad (7.4.40)$$

where eps is the desired allowable error. However, the usefulness of the iterative scheme is lost if the number of iterations required at a given time step become excessive.

7.4.5 The selection of time step size for heat and mass transfer and viscoelastic stress analysis

For this combined problem, five characteristic times are admissible. These are the time of heat diffusion, the time of mass diffusion, the relaxation times, the retardation times and the shifted time used in the stress analysis. These characteristic times are intrinsic properties of the material. Only four of them are independent because the relaxation times are related to the retardation times.

There are no relations between these characteristic times and the size of the time step to be used in the solution of the problem. For the particular material under consideration, the appropriate size of the time step has to be determined by trial and error. It should be noted that an unsuitable time step size will give inaccurate solutions for both heat and mass transfer and stress analysis. Since it would be computationally very expensive to retain a single step size for the full duration of the solution, the usual method is to start the solution process with a small time step and increase the size of the time step as the solution proceeds. This may be achieved automatically, by a simple check on convergence of the solution. If the solution does not converge to within the specified accuracy term, the time step size may be halved and the solution process continued as before.

7.4.6 Applications

7.4.6.1 Elastic analysis

The analysis of the elastic shrinkage behaviour of timber is shown for the case of a typical cross-section of $50 \times 100 \, \text{mm}^2$. The orientation of the grain was assumed to be that obtained by cutting the section from the heartwood of the tree, as shown in Figure 7.4.2. Symmetry of the problem allows analysis of only a quarter section. The finite element mesh, showing three regions of constant grain orientation to model the ring structure, is shown in Figure 7.4.3.

Details of the heat and masss transfer analysis of this problem are given in Reference 39. The elastic parameters used in the stress analysis are as follows:

$$E_x = 130 \, \text{Kips in}^{-2} \quad E_y = 125 \, \text{Kips in}^{-2} \quad E_z = 2310 \, \text{Kips in}^{-2}$$
$$G_{xy} = 16 \, \text{Kips in}^{-2}$$
$$v_{xy} = 0.556 \quad v_{yz} = 0.027 \quad v_{xz} = 0.285$$

The distinctly orthotropic nature of the problem is clearly represented in this data.

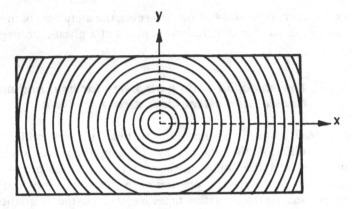

Figure 7.4.2 *Actual grain orientation—linear elastic orthotropic problem.*

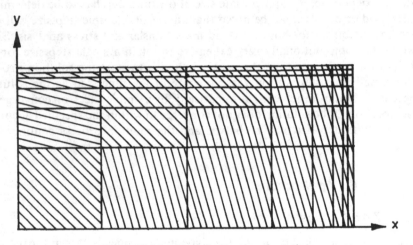

Figure 7.4.3 *Modelled grain orientation.*

Results of the analysis are presented in Figures 7.4.4 and 7.4.5 as contour plots of stresses. The hatched areas in Figure 7.4.4 represent zones where both principal stresses are compressive, whereas the hatched areas in Figure 7.4.5 signify zones where both principal stresses are tensile. High tensile stresses indicate critical areas for cracking and it can be seen therefore that these areas lie in the middle of the boundary edges. It was found that the shorter edge was the more critical location for all except very long drying times.

Over time, these maximum stresses develop in a characteristic fashion, as shown in Figure 7.4.6. From the unstressed initial state these critical stresses rise at a gradually lessening rate until the drying time is reached. After this time these stresses decrease at an exponential rate. The greatest values reached, at

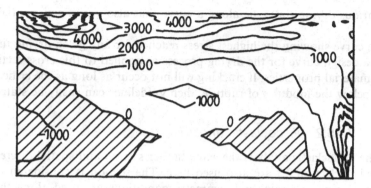

Figure 7.4.4   *Contour plot of major principal stresses.*

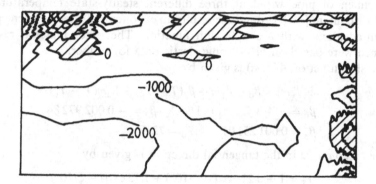

Figure 7.4.5   *Contour plot of minor principal stresses.*

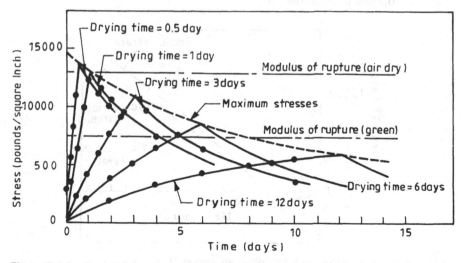

Figure 7.4.6   *Variation in maximum stresses at short edge midside during timber drying analysis.*

the various drying times, lie along a curve that also varies exponentially with time.

The curve showing the highest stress reached for any given drying time is in effect a design curve for the drying process as applied to this cross-section and these material properties. If cracking will not occur as long as the critical stress stays below the modules of rupture, then guidelines can easily be stated.

7.4.6.2 Viscoelastic analysis

For the viscoelastic analysis the same timber section and spatial discretisation as used in Section 7.3 are also used here. The section is considered to be in orthotropic plane strain, and symmetry conditions are used along the edges BC and AB. The material parameters used are based on creep experiments on a specimen of pine wood at three different steady-state temperatures and moisture content [40]. From these results an exponentially decaying creep curve was fitted along with a simple shift function. The other material responses assumed were based on orthotropic elastic data for timber.

The shift function $\Phi(T, m)$ is given by

$$\Phi(T, m) = \beta_0 + \beta_1 m + \beta_2(T - T_0) + \beta_3 m(T - T_0)$$
$$\beta_0 = -4/3 \qquad \beta_1 = 1/9 \qquad \beta_2 = -0.002\,972\,84$$
$$\beta_3 = 0.001\,251\,01 \qquad T_0 = 72°F$$

The creep curve in the tangential direction is given by

$$J_T = 1 + 0.25 \exp(-\xi/10^4) + 1.13 \exp(-\xi/10^7)$$

The compliances are evaluated by inverting the creep curve according to a method proposed by Srinatha [30].

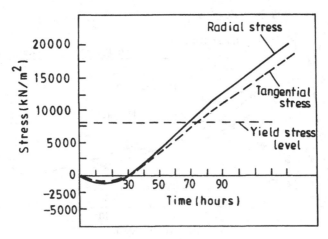

Figure 7.4.7 *Stresses at node 1 (centre).*

The tangential and radial stress distributions with time are shown in Figures 7.4.7–7.4.12 for six selected nodes: namely 1, 17, 32, 45, 47 and 69. The yield stress level for timber, beyond which the stresses enter the plastic region, is also shown in Figures 7.4.7–7.4.12. It is seen that at all these nodes the stresses are compressive during the initial stages of drying. After reaching a maximum value, the compressive stresses reduce in magnitude and a stress reversal takes place with the tensile stresses increasing in time.

At node 69, which is subjected to severe changes in moisture content, the stresses cross the yield stress level just after 2 h from the start of drying. At the

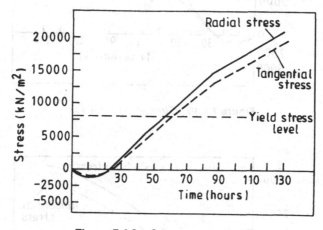

Figure 7.4.8 *Stresses at node 17.*

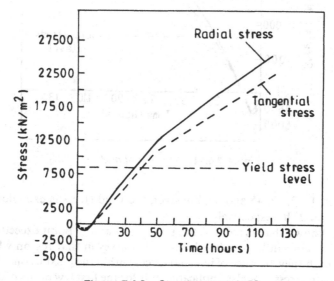

Figure 7.4.9 *Stresses at node 32.*

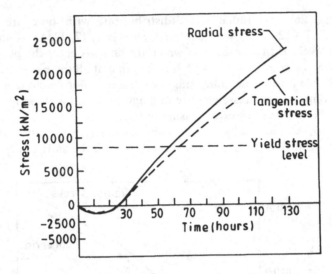

Figure 7.4.10   *Stresses at node 45.*

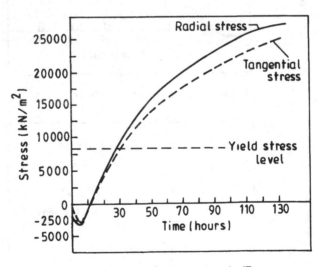

Figure 7.4.11   *Stresses at node 47.*

other nodes, 1, 17, 32, 45 and 47, however, the yield stress is exceeded after 68, 56, 36, 58 and 27 h, respectively.

It is well known that as soon as the stress level at any point exceeds the yield stress level, there will be a redistribution of stresses in that region which takes into account the new balance of internal forces within the plastic zone. Therefore, the viscoelastic stress model is applicable only for the first few hours of drying.

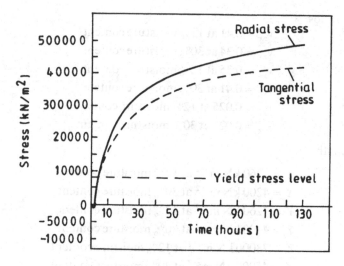

Figure 7.4.12   *Stresses at node 69.*

7.4.6.3   Elasto-plastic analysis

The elasto-viscoplastic rheological model has been used to determine the stress distribution in the plastic region for the period of time greater than 2 h from the start of the drying process. The same timber section and spatial discretisation as used in Section 7.3 are used again here.

To define completely the orthotropic behaviour of timber, 14 independent variables have to be specified: namely, the shrinkage coefficients, Young's moduli, Poisson's ratios, and yield strengths in the three principal directions, and the shear modulus and the shear yield strength in the radial–tangential plane. The numerical values used in this example have been quoted in Kollmann and Coté [41] and are given below:

Shrinkage coefficients:

$$(\alpha_s)_x = 0.00152\% \text{ moisture content}$$
$$(\alpha_s)_y = 0.00268\% \text{ moisture content}$$
$$(\alpha_s)_z = 0.00012\% \text{ moisture content}$$

Young's moduli:

$$E_x = 1.27 \times 10^6 \text{ kN m}^{-2} \text{ at } 12\% \text{ moisture content}$$
$$E_x = 0.92 \times 10^6 \text{ kN m}^{-2} \text{ at } 30\% \text{ moisture content}$$
$$E_y = 0.75 \times 10^6 \text{ kN m}^{-2} \text{ at } 12\% \text{ moisture content}$$
$$E_y = 0.55 \times 10^6 \text{ kN m}^{-2} \text{ at } 30\% \text{ moisture content}$$
$$E_z = 15 \times 10^6 \text{ kN m}^{-2} \text{ at } 12\% \text{ moisture content}$$
$$E_z = 12.3 \times 10^6 \text{ kN m}^{-2} \text{ at } 30\% \text{ moisture content}$$

Poisson's ratios:

$$v_{zx} = 0.41 \text{ at } 12\% \text{ moisture content}$$
$$v_{zx} = 0.34 \text{ at } 30\% \text{ moisture content}$$
$$v_{yx} = 0.35 \text{ at } 12\% \text{ moisture content}$$
$$v_{yx} = 0.41 \text{ at } 30\% \text{ moisture content}$$
$$v_{yz} = 0.025 \text{ at } 12\% \text{ moisture content}$$
$$v_{yz} = 0.021 \text{ at } 30\% \text{ moisture content}$$

Yield strengths:

$$X = 7260 \text{ kN m}^{-2} \text{ at } 12\% \text{ moisture content}$$
$$X = 4200 \text{ kN m}^{-2} \text{ at } 30\% \text{ moisture content}$$
$$Y = 7260 \text{ kN m}^{-2} \text{ at } 12\% \text{ moisture content}$$
$$Y = 4200 \text{ kN m}^{-2} \text{ at } 30\% \text{ moisture content}$$
$$Z = 78000 \text{ kN m}^{-2} \text{ at } 12\% \text{ moisture content}$$
$$Z = 45000 \text{ kN m}^{-2} \text{ at } 30\% \text{ moisture content}$$

Shear strength:

$$XY = 41000 \text{ kN m}^{-2} \text{ at } 8.8\% \text{ moisture content}$$
$$XY = 36000 \text{ kN m}^{-2} \text{ at } 15.6\% \text{ moisture content}$$

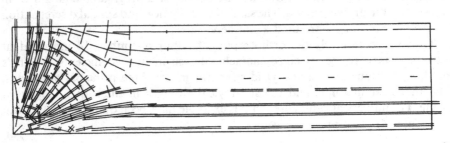

Figure 7.4.13 *Principal stress distributions at 2.4 h. Scale: 1 mm = 200 kN m^{-2}. Key: = tension; − compression.*

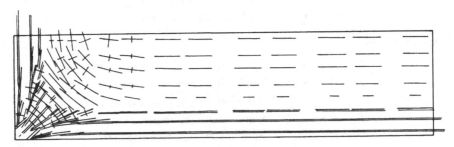

Figure 7.4.14 *Principal stress distributions at 9.5 h. For scale and key see Figure 7.4.13.*

Shear modulus:

$$G_{xy} = 1.68 \times 10^5 \, \text{kN m}^{-2} \text{ at } 12\% \text{ moisture content}$$
$$G_{xy} = 1.08 \times 10^5 \, \text{kN m}^{-2} \text{ at } 30\% \text{ moisture content}$$

The difference in values between tensile and compressive strengths is not taken into account. The rate of viscoplastic straining is evaluated using the same flow function as used in the isotropic Von Mises yield criterion.

Results of the analysis are shown in terms of plots of the principal stresses in the material at different stages of drying. Figures 7.4.13–7.4.16 show the principal stresses at 2.4, 9.5, 21.3 and 59.5 h of drying. The results obtained show a phenomenon which is of interest to the timber drying industry. It is known as 'stress reversal' and occurs when tensile stresses, initially established at the surface of the timber, change to compressive stresses and compressive stresses, initially established in the centre, become tensile. The changes take place because plastic deformations occur in the initial stages of drying, and as drying proceeds these deformations are locked into the material preventing further shrinkage taking place. The phenomenon can therefore only be modelled by means of plastic analysis. The results obtained clearly show this phenomenon taking place, lending support to the validity of the solution method employed.

The use of an orthotropic analysis enables the increased strength and stiffness of the material in the longitudinal direction to be included in the analysis and

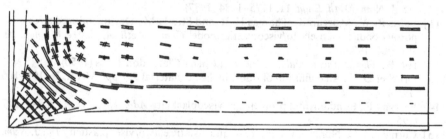

Figure 7.4.15   *Principal stress distributions at 21.3 h. For scale and key see Figure 7.4.13.*

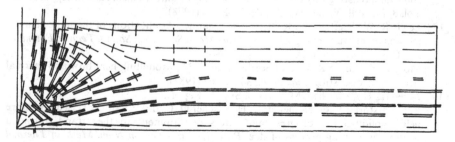

Figure 7.4.16   *Principal stress distributions at 59.5 h. For scale and key see Figure 7.4.13.*

it is this component that provides the main support for carrying the high drying stresses induced.

REFERENCES

[1] Zienkiewicz O. C. and Taylor R. L., *The Finite Element Method, Volume 1: Basic Formulations and Linear Problems* (4th Edition) McGraw-Hill (1988).

[2] Lewis R. W. and Bass B. R., The determination of stresses and temperatures in cooling bodies by finite elements *Trans. ASME, J. Heat Transfer*, **98**, 478–484 (1976).

[3] Williams J. R., Lewis R. W. and Morgan K., An elasto-viscoplastic thermal stress model with applications to the continuous casting of metals *Int. J. Num. Meth. Engg*, **14**, 1–9 (1979).

[4] Zienkiewicz O. C. and Cormeau I. C., Viscoplasticity–plasticity and creep in elastic solids—a unified numerical solution approach *Int. J. Num. Meth. Engg*, **8**, 821–845 (1974)

[5] King S. Y. and Halfter N. A., *Underground Power Cables* Longman (1982).

[6] Krischer O., *Die Wissenschaftlichen Grundlagen der Trocknungstechnik* Springer (1963).

[7] Philip J. R. and de Vries D. A., Moisture movement in porous materials under temperature gradients *Trans. Am. Geophys. Union*, **38**, 222–232 (1957).

[8] de Vries D. A., Transfer of heat and moisture in porous media *Trans. Am. Geophys. Union*, **39**, 909–916 (1958).

[9] Luikov A. V., *Heat and Mass Transfer in Capillary Porous Bodies* Pergamon (1966).

[10] Lewis, R. W., Strada M. and Comini G., Drying induced stresses in porous bodies *Int. J. Num. Meth. Engg*, **11**, 1175–1184 (1977).

[11] Lewis R. W., Morgan K., Thomas H. R. and Strada M., Drying induced stresses in porous bodies—an elasto/viscoplastic model *Comp. Meth. Appl. Mech. Engg*, **20**, 291–301 (1979).

[12] Hill R., *The Mathematical Theory of Plasticity* Clarendon Press (1950).

[13] Drucker D. C., An definition of stable inelastic material *J. Appl. Mech.*, **26**, 101–106 (1959).

[14] Perzyna P., Fundamental problems in viscoplasticity *Adv. Appl. Mech.*, **9**, 243–377 (1966).

[15] Cormeau I. C., Numerical stability in quasi-static elasto/viscoplasticity *Int. J. Num. Meth. Engg*, **9**, 109–127 (1975).

[16] Morgan K., Lewis R. W. and Williams J. R., Thermal stress analysis of a novel continuous casting process *The Mathematics of Finite Elements and its Applications*, Vol. 3, Ed. J. R. Whiteman, Academic Press (1978).

[17] Lewis R. W., Seetharamu K. N. and Morgan K., Applications of the finite element method in the study of ingot casting *Proc. Metal Soc. Conf. Solidif. Tech. in Foundary and Cast House* Warwick, pp. 40–43 (1980).

[18] Morgan K., Lewis R. W. and Seetharamu K. N., Modelling heat flow and thermal stress in ingot casting *Simulation*, **36**, 55–63 (1981).

[19] Jones H., Unpublished, Ministry of Supply Report.

[20] Rammerstorfer F. G., Jaquemar C. H., Fischer D. F. and Wiesinger H., Temperature fields, solidification progress and stress development in the strand during a continuous casting process of steel *Proc. 1st Int. Conf. on Num. Meth. in Thermal Problems* Swansea, pp. 712–722 (1979).

[21] Rossi I., *Patent Specification, 1-400-811* (1972).

[22] Onsager L., Reciprocal relations in irreversible processes *Phys. Rev.* **37**, 405–426 (1931).

[23] Comini G. and Lewis R. W., A numerical solution of two dimensional problems involving heat and mass transfer *Int. J. Heat Mass Transfer*, **19**, 1387–1392 (1976).

[24] Thomas H. R., *A Finite Element Analysis of Shrinkage Stresses in Building Materials* Ph.D. Thesis, University of Wales (1980).

[25] Thomas H. R., Lewis R. W. and Morgan K., An application of the finite element method to the drying of timber *Wood and Fibre*, **11**, 123–130 (1979).

[26] Thomas H. R., Morgan K. and Lewis R. W., A fully non-linear analysis of heat and mass transfer problems *Int. J. Num. Meth. Engg*, **15**, 1381–1393 (1980).

[27] Lewis R. W., Morgan K. and Thomas H. R., A non-linear analysis of shrinkage stresses in porous materials *Proc. 1st Int. Conf. on Num. Meth. in Thermal Problems* Swansea, pp. 515–526 (1979).

[28] Lewis R. W., Morgan K. and Thomas H. R., The non-linear modelling of drying-induced stresses in porous bodies *Advances in Drying*, Vol. 2 Hemisphere Publishing Corporation, pp. 233–268 (1983).

[29] Hill R. Anisotropic plasticity *Proc. Roy. Soc.*, **193**, 281–292 (1948).

[30] Srinatha H. R., *A Finite Element Method for the Stress Analysis of Plane Thermovisco-elastic Bodies* Ph.D. Thesis, University of Wales (1981).

[31] Srinatha H. R. and Lewis R. W., A finite element method for thermoviscoelastic analysis of plane problems *Comp. Meth. Appl. Mech. Engg*, **25**, 21–33 (1981).

[32] Hopkins I. L. and Hamming R. W., On creep and relaxation *J. Appl. Phys.*, **28**, 906–917 (1957).

[33] Lee E. H. and Rogers T. G., Solution of viscoelastic stress analysis problems using measured creep or relaxation functions *J. Appl. Mech.*, **30**, 127–133 (1963).

[34] Morgan K., Lewis R. W. and Thomas H. R., Numerical modelling of drying induced stresses in porous bodies *Developments in Drying* Science Press, New York (1979).

[35] Morgan K., Lewis R. W. and Thomas H. R., Finite element modelling of drying stresses in timber and cooling stresses in cast metal *Modelling and Simulation in Practice*, **2**, 51–76 (1979).

[36] Lewis R. W., Morgan K. and Thomas H. R., The determination of hygrothermal stress development in anisotropic composites *Proc. 2nd. Int. Conf. on Num. Meth. in Thermal Problems* Venice, pp. 301–312 (1981).

[37] Morgan K., Thomas H. R. and Lewis R. W., The numerical modelling of stress reversal in timber drying *Wood Sci.*, **15**, 139–149 (1982).

[38] Srinatha H. R. and Lewis R. W., A finite element formulation of uncoupled thermo-viscoelastic response of plane problems for all admissible values of Poisson's ratio *Int. J. Num. Meth. Engg*, **18**, 765–774 (1982).

[39] Johnson K. H., White I. R., Lewis R. W. and Morgan K., The analysis of drying-induced stresses in wood *Proc. 2nd. Int. Conf. Appl. Num. Modelling* Madrid (1978).

[40] Johnson J. A., Private communication (1981).

[41] Kollmann F. F. P. and Coté W. A. J., *Principles of Wood Science and Technology*, Vol. 1: *Solid Wood* Springer-Verlag, Berlin (1968).

[42] Ferguson W. J., *A Finite element model for heat and mass transfer in capillary-porous bodies with particular reference to the influence of pressure gradient* Ph.D. Thesis, University of Wales (1991).

[43] Lewis R. W. and Ferguson W. J., The effect of temperature and total gas pressure on the moisture content in a capillary-porous body *Int. J. Num. Meth. Engg*, **29**, 357–369 (1990).

[44] Lewis R. W., Sze W. K. and Roberts P. M. A finite element investigation into heat and mass transfer in porous materials with particular reference to ground freezing, Keynote Lecture presented at *Drying, 86* Boston, August (1986), Also, *Drying 86*, Vol. 1, pp. 22–29 Hemisphere Publishing Company.

[45] Lewis R. W. and Sze W. K., A finite element simulation of frost heave in soils *5th Int. Symp. on Ground Freezing*, Eds. Jones and Holden, pp. 73–80 Balkena (1988).

[46] Lewis R. W., Sze W. K. and Huang H. C., Some novel techniques for the finite element analysis of heat and mass transfer problems *Int. J. Num. Meth. Engg*, **25**, 611–624 (1988).

[47] Ferguson W. J., Lewis R. W. and Tömösy L., *A Finite Element Analysis of Freeze-Drying of a Coffee Sample* Report No. CR/694/91 Department of Civil Engg, Univ. College of Swansea (1991).

Nomenclature

a_m	Mass transfer coefficient
C_m	Specific mass capacity
C_p	Specific heat capacity
C_d	Defined by equation (7.4.6)
C_{dz}	Defined by equation (7.4.9)
C_m	Generalised mass capacity—see equation (7.3.35)
C_n	E_x/E_y
C_{nz}	E_z/E_y
C_q	Generalised heat capacity—see equation (7.3.35)
C_{ij}	Material compliance functions
d_m	Small change in moisture content
$d_e m$	Small change in moisture content due to moisture transfer
$d_i m$	Small change in moisture content due to phase conversion
E	Young's Modulus
F	Yield function
F_0	Reference value of yield function
G	Shear modulus
h	Specific enthalpy
T	Volumetric capacity of the source or sink of moisture dependent on phase change
j_i	Flux
j_m	Moisture flux
j_q	Heat flux
j_1	Vapour flux
j_2	Liquid flux
J_m^*	Generalised moisture flux—see equation (7.3.45)
J_q^*	Generalised heat flux—see equation (7.3.46)
J_T	Creep curve in the tangential direction
J_1, J_2, J_3	Stress invariants
J_1', J_2', J_3'	Stress invariants in terms deviatoric stresses
k	Coefficient of thermal conductivity
k_m	Coefficient of mass conductivity
k_{il}	Conductivities—general
k'	Defined by equation (7.2.18)

K	Hardening parameter
K_m, K_q, K_E, K_δ	Generalised conductivity—see equation (7.3.35)
l	Length
L	Latent heat of fusion
L_{il}	Phenomenological coefficients
m	Mass or moisture content
m_a	Ambient moisture content
m_w	Specified moisture content at the boundary
m_0	Reference moisture content
N_r, N_i, N_s	Shape functions
Q	Plastic potentials
s	Entropy
t	Time
t'	Creep time—see equation (7.4.19)
Δt	Time step size
T	Temperature
T_a	Ambient temperature
T_c	Temperature causing solidification
T_i	Initial temperature
T_s	Nodal values of temperature
T_w	Specified temperature at the boundary
T_0	Reference temperature
$\hat{T}$	Approximate value of temperature
u	Moisture potential
u_a	Ambient moisture potential
u_s	Nodal values of moisture potential
u_w	Specified moisture potential at the boundary
u	Approximate value of moisture potential
U'	Internal energy
x	Co-ordinate direction
X	Tensile yield stress in the x-direction
XY	Yield stress in shear on the xy-plane
y	Co-ordinate direction
Y	Tensile yield stress in the y-direction
YZ	Yield stress in shear on the yz-plane
z	Co-ordinate direction
Z	Tensile yield stress in the z-direction
ZX	Yield stress in shear on the zx-plane
α_m	Convective mass transfer coefficient in terms of moisture content
α_q	Convection heat transfer coefficient
α_s	Shrinkage coefficient
α_T	Coefficient of thermal expansion
α_u	Convective mass transfer coefficient in terms of moisture potential
β	dm_1/dm_2

v_{xy}	Shear strain on the xy-plane
v'	Fluidity parameter
Γ	Boundary
Γ_1	Part of the boundary subject to constant temperature
Γ_2	Part of the boundary subject to heat flux conditions
Γ_3	Part of the boundary subject to constant moisture content
Γ_4	Part of the boundary subject to moisture flux conditions
δ	Displacement
δ_r	Nodal displacement
$\hat{\delta}$	Approximate value of displacement
δ_{ij}	Kronecker delta
$\delta*$	Thermogradient coefficient
δ'	$\delta*/c_m$
ε_{ij}	Strain
ε_{ij}^e	Elastic strain
ε_{ij}^{vp}	Viscoplastic strain
ε_{ij}^i	Initial strain
ε_{ij}^p	Plastic strain
ε'	Phase conversion factor
ζ_i	Relaxation times
θ_i	Defined by equation (7.3.1)
K_1-K_6	Parameters characteristic of the state of anisotropy
λ	Latent heat of vaporisation
$d\lambda$	Positive constant of proportionality—see equation (7.2.8)
μ	Chemical potential
v_{ij}	Poisson's ratio
ξ	Shifted time
ξ'	Shifted time prior to ξ
π_i	Parameter which may be pressure, temperature, concentration, etc.
π_i^*	Deviation of π_i from its equilibrium value
ρ	Density
σ_{ij}'	Defined by equation (7.2.17)
σ_{ij}	Stress
σ	$(J_2')^{1/2}$
τ_{ij}	Shear stress
ϕ_i	Thermodynamic forces
Φ	Shifting function
$\Psi(F)$	Flow function
$\Omega(F/F_0)$	Flow function/fluidity parameter
Ω	Domain of interest
b	Body forces
B	Matrix of derivatives of shape function
C(Φ)	General capacity matrix
C$_m$	Mass capacity matrix

$\mathbf{C}_q$	Heat capacity matrix
$\mathbf{D}$	Elasticity matrix
$\mathbf{F}^e$	Element force vector—see equation (7.4.25)
$\mathbf{F}^e_{m1}$	Defined by equation (7.4.35)
$\mathbf{F}^e_{ML}$	Vector of memory load due to strains—see equation (7.4.38)
$\mathbf{F}^e_{MTm}$	Vector of memory loads due to changes in temperature and moisture content—see equation (7.4.38)
$\mathbf{F}^e_{T}$	Total load vector
$\mathbf{F}^e_{Tm}$	Load vector due to changes in temperature and moisture content—see equation (7.4.27)
$\mathbf{J}(\Phi)$	Vector of generalised fluxes
$\mathbf{J}_m$	Vector of generalised mass fluxes
$\mathbf{J}_q$	Vector of generalised heat fluxes
$\mathbf{K}$	Overall stiffness matrix in stress analysis
$\mathbf{K}(\Phi)$	Overall generalised conductivities matrix
$\mathbf{K}_m$	Generalised mass conductivity matrix
$\mathbf{K}_q$	Generalised thermal conductivity matrix
$\mathbf{K}1$	Initial elastic stiffness matrix in the viscoelastic analysis—see equation (7.4.26)
$\mathbf{K}2$	Defined by equation (7.4.34)
$\mathbf{K}_\varepsilon, \mathbf{K}_\delta$	Generalised coupling matrices in heat mass transfer analysis—see equation (7.3.52)
$\mathbf{N}$	Shape functions
$\mathbf{q}$	Defined by equation (7.4.37)
$\mathbf{R}$	Total load vector in the thermal stress analysis
R_b	Defined by equation (7.2.32)
$\mathbf{T}$	Vector of nodal temperatures
$\mathbf{t}$	Prescribed surface tractions
$\mathbf{u}$	Vector of nodal moisture potentials
$\boldsymbol{\delta}$	Vector of nodal displacements
$\boldsymbol{\varepsilon}$	Strain vector
$\boldsymbol{\varepsilon}^e$	Elastic strain vector
$\boldsymbol{\varepsilon}^i$	Initial strain vector
$\boldsymbol{\varepsilon}^{vp}$	Visco plastic strain vector
$\boldsymbol{\varepsilon}_m$	Initial strain vector due to moisture content changes
$\boldsymbol{\varepsilon}_T$	Initial strain vector due to temperature change
$\boldsymbol{\varepsilon}^*$	Defined by equation (7.4.31)
$\boldsymbol{\sigma}$	Stress vector
$\boldsymbol{\sigma}^e$	Elastic stresses vector
$\boldsymbol{\sigma}^i$	Initial stresses vector
$\boldsymbol{\phi}$	Vector of all nodal unknowns in the heat and mass transfer analysis

Index

Index compiled by Geoffrey C. Jones